T0257534

Handbook of Protein Engineering

Handbook of Protein Engineering

Edited by **Anton Torres**

New York

Published by Callisto Reference,
106 Park Avenue, Suite 200,
New York, NY 10016, USA
www.callistoreference.com

Handbook of Protein Engineering
Edited by Anton Torres

International Standard Book Number: 978-1-63239-410-1 (Hardback)

Printed in the United States of America.

Contents

Preface

It is often said that books are a boon to mankind. They document every progress and pass on the knowledge from one generation to the other. They play a crucial role in our lives. Thus I was both excited and nervous while editing this book. I was pleased by the thought of being able to make a mark but I was also nervous to do it right because the future of students depends upon it. Hence, I took a few months to research further into the discipline, revise my knowledge and also explore some more aspects. Post this process, I begun with the editing of this book.

The field of protein engineering has been comprehensively illustrated in this book. The aim of this book is to provide state-of-the-art information regarding the field of protein engineering and elucidate its applications as well as technology. It covers a broad spectrum of significant topics like chromatography methodology, protein-protein and protein-ligand docking, protein engineering of enzymes involved in bio-plastic metabolism, etc. It will appeal to a wide range of readers including researchers, scientists, and even students who wish to gain knowledge about the principles and practices of protein engineering.

I thank my publisher with all my heart for considering me worthy of this unparalleled opportunity and for showing unwavering faith in my skills. I would also like to thank the editorial team who worked closely with me at every step and contributed immensely towards the successful completion of this book. Last but not the least, I wish to thank my friends and colleagues for their support.

Editor

Basic Technology

Chromatography Method

Jingjing Li, Wei Han and Yan Yu

Additional information is available at the end of the chapter

1. Introduction

Term 'chromatography' was firstly employed by Russian Scientist Mikhail Tsvet in 1900 to describe the phenomenon that a mixture of pigments was carried by a solvent to move on paper and separated from each other. Since the pigments have different colors, the phenomenon was the termed by "chromato-graphy' literally means 'color writing' [1]. Now, it is generally refers to a series techniques for the separation of mixtures [2].

Each chromatography involves two phases, mobile phase and stationary phase. The mobile phase drives compounds to flow through the surface of the stationary phase and the movements of compounds are retarded by interaction with stationary phase. Compounds are retarded differentially according to the strength the interaction and finally are separated.

The chromatography was early performed on papers or thin layers to separate small molecule compounds, termed planar chromatography (Figure 1A). Later, the column chromatography was developed, in which the stationary phase is manufactured into porous particle media and parked in a column and the mobile phase flows through thin channels among media [3]. If the mobile phase is gas and stationary phase is liquids, the technique is termed gas chromatography [4], which is used in separation of volatile compounds (Figure 1B). If the mobile phase is liquid and stationary phase is solid, it is termed liquid chromatography [5] and used widely in separation of small compounds or biological macromolecules (Figure 1C).

The liquid chromatography is the most popular technique in protein purification and analysis. The liquid mobile phase containing proteins flows through the column and is separated by interacted with media. The stationary phase composed by porous particles supplies much more surface compared with traditional planar chromatography. So the loading capacity is much more increased and could purify even grams of protein in one cycle. Furthermore, the column structure provides a possibility to employ high pressure to drive the mobile phase flowing much faster and complete a separation within short time, termed high pressure liquid

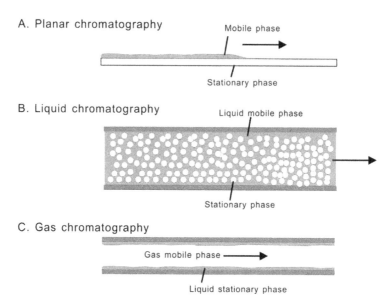

A. Planar chromatography

Mobile phase

Stationary phase

B. Liquid chromatography

Liquid mobile phase

Stationary phase

C. Gas chromatography

Gas mobile phase

Liquid stationary phase

Figure 1. Different types of chromatography

chromatography [6]. At same time, uniform size of matrix benefited by exquisite quality gives the column chromatography much higher resolution than before. The high performance in high loading capacity, high flow rate, and high resolution made the column chromatography become the most rapidly developed protein separation technique in the last two decades.

Several basic types of chromatography had been developed based on different separation properties (Table 1). This chapter describes both principles and applications of these techniques.

Property	Technique
Net charge	Ion exchange chromatography
Hydrophobicity	Hydrophobic interaction chromatography and
	Reverse phase chromatography
Biorecognition	Affinity chromatography
Size	Size exclusion chromatography

Table 1. Different chromatography techniques and corresponding protein properties

2. Ion-exchange chromatography

Ion-exchange chromatography (IEXC) was introduced to protein separation in the 1960s and plays a major role in the purification of biomolecules [7]. IEXC separation is based on the reversible electrostatic interactions between charged solutes and an oppositely charged medium. The technique is straightforward on its theory and operation, so that easily to be grasped by beginners.

Ion exchange refers to the exchange of ions between two electrolytes or between an electrolyte solution and a complex. For example: $NiSO_4 + Ca^{2+} = CaSO_4 + Ni^{2+}$. When one of the electrolytes was immobilized on resin, the exchange will happen between the interface of liquid phase and solid phase, termed exchanger, such as,

$$R\text{-}O\text{-}CH_2\text{-}COOY + X^+ \rightarrow R\text{-}O\text{-}CH_2\text{-}COOX + Y^+ \tag{1}$$

In which R indicates the base matrix portion of the resin, the ion X^+ exchanges with Y^+ and is adsorbed by resin.

The exchange reaction is reversible and the direction depends on the concentration and ionization constant of the electrolytes. In Equation 1, if concentration of ion Y^+ increases, X^+ will be desorbed.

$$R\text{-}O\text{-}CH2\text{-}COOX + Y^+ \rightarrow R\text{-}O\text{-}CH2\text{-}COOY + X^+ \tag{2}$$

The ion Y^+ could be any cation, such as Na^+, H^+. The two equations present the process of the binding and elution in IEXC.

According to the above two equations we know the binding of protein on exchanger is a kinetic equilibrium between adsorption and desorption. The equilibrium constant Kd is:

$$Kd = \left[X^+\right]\left[R\text{-}O\text{-}CH2\text{-}COOY\right]/\left[Y^+\right]\left[R\text{-}O\text{-}CH2\text{-}COOX\right] \tag{3}$$

With mobile phase moving, protein molecules in mobile phase are carried forward and adsorbed by downstream medium, at same time adsorbed proteins are released from stationary phase to mobile phase. Proteins remove forward companied with continuous adsorption and desorption. Under the same ionic strength, the higher Kd a protein has, the more fraction distributes in mobile phase and moves faster. Reversely, proteins having smaller Kd are more retarded than that having larger Kd. Actually, all kinds of adsorption chromatographys are base on the kinetic equilibrium mechanism.

2.1. Isoelectric point of protein

Proteins are ampolytes on which carboxyl groups and amino groups of side chains and two terminals could ionize and cause proteins being positively and negatively charged. The positive charges of proteins typically attribute to ionized cysteine, aspartate, lysines, and histidines. Negative charges are principally provided by aspartate and glutamate residues. At a certain pH point, the total positive charges of a protein equal to the total negative charges, the net charge is 0 at this time and the pH is defined as the isoelectric point (pI) of this protein. When the solution pH higher than pI of a protein, more carboxyl groups ionized and the protein is negatively charged, vise versa (Figure 2). pI of a protein could be determined by several experiment methods, but an approximate value could be calculated by mathematics methods. Once a protein primary structure is given, the pI can be calculated by software or some concise websites such as:

http://web.expasy.org/compute_pi/

http://www.scripps.edu/~cdputnam/protcalc.html

2.2. Selection of exchanger

The exchangers in IEXC are composed of base matrix and functional groups that coupled on surface of the matrix. The base matrix is nonporous or porous spherical particles with charge free surface on which different functional groups link. Porous matrix offers a large surface area for protein binding and so gives a high binding capacity, but sacrificed some resolution due to the diffusion between outside and inside of matrix. On the contrary, nonporous matrix is limited on binding capacity, but used to provide high resolution on micropreparative or analytical separations.

Similar to the effect of porosity, the size of particles also influences the resolution of all kinds of chromatography including IEXC. Even and small particle size facilitates the efficient transfer of molecules between the mobile and the stationary phases, and provides high resolution, but increases the resistance of the column so that needs higher pressure or longer separation time. Small size particles are preferable for analytical separations. On the contrary the large size particles are more used on large scale production.

The selectivity of ion exchange media depends briefly on the nature and substitution degree of the functional groups, or called ligands. The media are classified into anion exchangers and cation exchangers. Ligand of the anion exchangers can be positively charged and anions can bind and exchange on it. On the contrary, the cation exchangers can be negatively charged on which cations exchange. The commonly used exchangers named after the functional groups and list in Table 2.

Exchanger	Ligand	Charged group
Strong cation	Sulfopropyl (SP)	$-CH_2CH_2CH_2SO_3^-$
Weak cation	Carboxymethyl(CM)	$-O-CH_2COO^-$
Strong anion	Quaternary ammonium (Q)	$-N^+(CH_3)_3$
Weak anion	Diethylaminoethyl (DEAE)	$-N^+H(C_2H_5)_2$
	Diethylaminopropyl (ANX)	$-N^+H(C_2H_5)_2$

Table 2. Commonly used exchangers

Ion exchangers are classified as weak or strong according to the ionization properties of ligands. The strong exchangers own ligands with high ionization coefficient (Figure 2). They are fully charged in pH range 1~13. In this range, pH change does not influence the charge of the ion exchanger. Thus the strong exchangers can wide used in almost all pH range. On the contrary, weak ion exchangers have weak electrolytes as functional ionic groups. The ionization of these groups is influenced by solution pH. So that they can offer a different selectivity compared to strong ion exchangers.

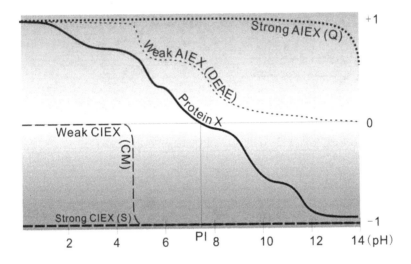

Figure 2. Charge property of the common types of ion exchangers and example protein with different pH value. (Modified from Ion Exchange chromatography & chromatofocusing, principle and methods, GE healthcare)

2.3. Surface charge of protein

The mobile phase in IEXC is aqueous solution with proper pH value and ionic strength. The pH value determines the charge property of protein. A pH value lower than a protein pI will causes a positive net charge of the protein and vise versa. It should be noted that IEXC is base on the electrostatic interaction. The interaction between a protein and an ion exchanger depends more on the charge distribution of the protein surface than the net charge (Figure 3). The distribution of the charge on surface and internal is not even, so a solution with pH value slightly different to protein pI could not insure the protein exhibit an expected charged surface. In practice, the pH is typically at least 1 unit higher or lower than pI of target protein to ensure the protein has an expected surface charge.

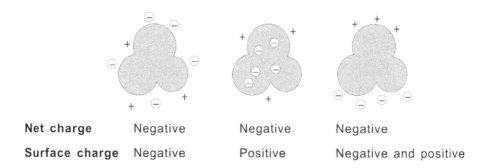

Net charge	Negative	Negative	Negative
Surface charge	Negative	Positive	Negative and positive

Figure 3. Different charge distributions of proteins.

2.4. Mobile phase

Mobile phase is composed of pH buffer system and neutral salt ions. Buffering ions in buffer should have the same charge with exchanger. Otherwise the buffering ions will bind to exchanger prior to eluent ions and cause significant pH fluctuation during elution. The commonly used buffers are given in table 3.

Buffers	pH range at 20 mM
Buffer for cation exchange chromatography	
Citric acid	2.6~3.6
Acetic acid	5.3~6.3
MES	5.8~6.8
Phosphate buffer	6.3~7.3
HEPES	7.1~8.1
Buffer for anion exchange chromatography	
Bis-tris	6.0~7.0
Tris-HCl	7.5~8.5
TEA	7.4-8.8
Ethanolamine	9.0~10.0
Piperdine	10.5~11.5

Table 3. Commonly used buffer for cation and anion exchange chromatography

Except pH value, the ionic strength also influences the binding of the protein. A typical IEXC experiment includes a binding stage and an elution stage. As indicated in Equation 1 and 2, proteins tend to be adsorbed by exchanger at low ionic strength and be desorbed at high ionic strength. So the ionic strength should be low enough in binding process to ensure protein adsorption and increased to elute proteins. The ionic strength in IEXC is usually modulated by adding high concentration of NaCl solution.

2.5. Operation

2.5.1. Binding process

All solutions used in column chromatography, including sample solution, should be degased and filtered (0.22 or 0.45 um membrane) to avoid the clogging of column by air bubbles or particles. Before sample loading the column should be equilibrated with 2 column volumes (CV) of initial buffer. And then sample is loaded with same flow rate. After that 3 CV of initial buffer should be run to wash off the unbound impurity proteins.

2.5.2. Elution

Although proteins could be separated under constant solvent composition, termed isocratic elution, for most tightly adsorbed proteins, it will take very long time to be eluted.

In practice, the mostly used strategy is to accelerate the exchange of protein by increasing ion strength in initial buffer. The most widely used agent is NaCl. It is convenient to increase the cation Na^+ and anion Cl^- at same time and without significantly change pH value of solution. Proteins could be eluted by linear or stepwise gradient ion strength or combination of them (Figure 4). The stepwise gradient elution is used in group separation. In each step one group of proteins with similar charge property is eluted simultaneously. It is often used in large scale production. While, linear gradient could be seem as infinite number of tiny steps, in which protein was eluted and separated one by one. It is more used in preliminary experiments or analytical separations. In practice, the usual strategy is combination of linear and stepwise gradient. As show in figure 4C, a part of impurities are eluted first by a step elution, and then the target protein is separated from the similar charged protein by a linear elution.

Another elution method is to change the surface charge of proteins by changing pH value of the elution buffer. Typically, in cation IEXC, increased pH value decreases the surface positive charge and the interaction between proteins and exchangers is weakened. Reversely the pH value is decreased in anion IEXC to elute protein. Proteins are eluted at the pH value close to their pI. It should be noted that, change of pH could also alter the charge property of weak exchangers in certain ranges, so the weak exchanger possibly gives different resolution in these ranges. But pH elution is less used in practice because some proteins precipitate at pH value near to their pI and clog column. Additionally, it is hard to keep ion strength constant as changing pH value and present a worse reproducibility.

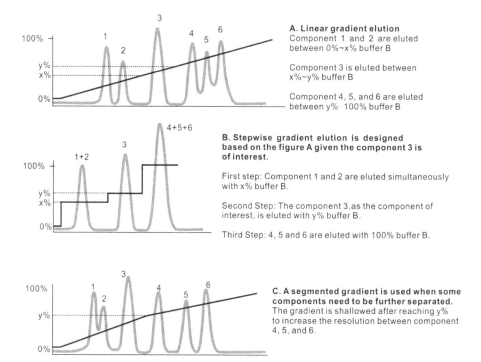

Figure 4. Different strategies of gradient elution.

2.6. Feature and application

IEXC is one of the most frequently used chromatographic techniques for the protein separation. The adsorption and elution take place under mild condition so that the natural activities can be well maintained during chromatographic process.

2.6.1. Purification of recombinant human Midkine by SP column

A recombinant human Midkine, pI=9.7, was expressed by a yeast fermentation technology and separated by IEX chromatography using SP column. The fermentation culture with high potassium phosphate buffer (100 mM) was diluted by pure water, lowering the conductivity to <10 mS/cm, and adjusted to pH 6.2 by Na_2HPO_4 solution. 50 ml Sepharose FF column with maximum loading capacity of 70 mg/ml was used to capture total 200 mg proteins in sample solution. A fraction of non-target protein was eluted by stepwise elution using 0.5 M NaCl, and then a linear gradient from 0.5~1.0 M NaCl was used to separated the target protein from the other impurities.

Sample: 6000 ml Yeast X33 fermentation culture containing
recombinant human Midkine (~0.04mg/ml)
Column: SP FF column (50 ml)
Buffer A: 20 mM sodium phosphate, pH 6.2
Buffer B: 20 mM sodium phosphate, 2M NaCl, pH 6.2
Gradient: 25~50%B in 200 ml(6CV) where 50%B=1.0 M NaCl
Flow rate: 5 ml/min

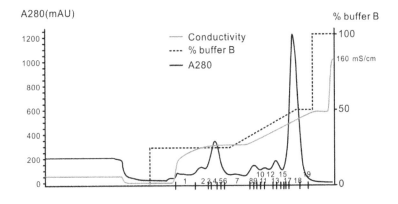

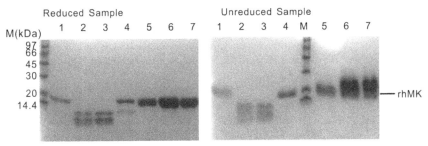

Unreduced Sample
Lane 1: Crude sample from fermentation medium
Lane 2: fraction 4
Lane 3: fraction 5
Lane 4: fraction 13
Lane 5: fraction 15
Lane 6: fraction 18
Lane 7: fraction 19

Figure 5. Cation IEXC of rhMK (result of Shixiang Jia, Ping Tu et al. General regeneratives (shanghai) limited, Shanghai, PR China)

3. Hydrophobic interaction chromatography

Hydrophobic interaction chromatography (HIC) bases on the interactions between hydrophobic surface of proteins and hydrophobic ligands on the medium [8]. It is used in protein separation for more than a half century, although there is not a widely accepted theory to define the hydrophobic interaction.

The principle of HIC is parallel to that of salting out. In aqueous solution, hydrogen bond is formed between water molecules and protein surface. By hydrogen bond, the side chains of protein molecules adsorb water molecules to form an ordered water film around them. The water film prevents protein molecules from aggregating and precipitating. Different amino acid side chains have variant abilities in forming hydrogen bond. Hydrophobic amino acids, such as isoleucine, valine, leucine, and phenylalanine, tend to loss their ordered water as solution ion strength increases. Relative hydrophobicity of amino acids was defined by the change of Gibbs free energy when amino acids are transferred from aqueous solution to non-polar solvent [9]. The distribution of hydrophobic amino acids on protein surface determines the hydrophobicity of the protein. As salt concentration increases, proteins associate each other and precipitate in the order of decreasing hydrophobicity. This process is termed fractional salting out (Figure 6B).

In HIC, the concentration of salt is controlled at an appropriate value, for example, 1 M $(NH_4)_2SO_4$. At this concentration, the hydrophobic interaction still not strong enough to cause proteins precipitate. However, the hydrophobic media, termed adsorbent, could adsorb proteins by high hydrophobic ligand coupled on it (Figure 6C). When protein solution flows through the HIC column, proteins having certein hydrophobicity will be adsorbed, and proteins with weak hydrophobicity will flow through with mobile phase. So, to adsorb proteins with weak hydrophobicity needs application of higher salt concentration or medium with stronger hydrophobicity to increase the hydrophobic interaction.

3.1. Stationary phase

The media of HIC are composed of base matrix and ligand. Base matrix functions as a support on which the hydrophobic ligand is immobilized. To avoid the disturbance the hydrophobic interactions between proteins and ligand, the matrix should have an inert surface. Cross-linked agarose is one of the most widely used matrix, it has a porous structure, having high binding capacity, high flow rate, good physical and chemical stability. Except that silico or synthetic copolymer materials are also widely used matrix.

Hydrophobic ligands are attached to the surface of base matrix by covalent bonds, for example, by glycidyl-ether for agarose and silyl-ether for silico gel. Widely used ligands for HIC are linear chain alkanes and phenyl. The strength of the hydrophobicity increases with the increase of length of the carbon chain. Butyl (C4) and octyl (C8) are often used linear chain ligands. Another widely used ligand is phenyl, which not only has a same hydrophobicity with pentyl ligand, also has a potential for π-π interactions with proteins rich in aromatic groups.

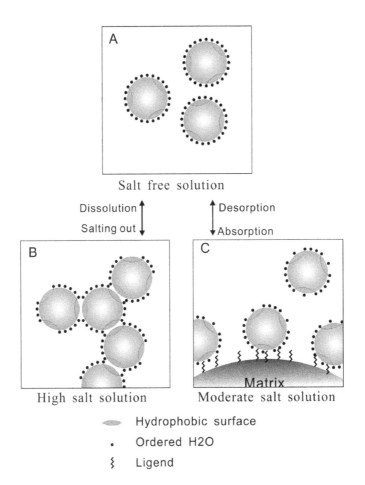

Figure 6. Salting out process and adsorption between protein and adsorbent. (A) A protein can disperse in salt free solution. (B) When salt concentration increases, the ordered water molecules are taken up. Proteins tend to aggregates and precipitates. (C) With a moderate salt concentration, the hydrophobic interaction between protein molecules is not strong enough to cause salting out, but can result in proteins adsorbed by hydrophobic matrix.

Before separating of each new protein, it is a good idea to screen different media by pretests on small prepacked column. The pretests should start from the medium with lowest hydrophobic. An ideal medium should firstly have an appropriate hydrophobicity by which the target protein could be adsorbed at a certain salt concentration. The lower hydrophobic a protein is, the higher hydrophobicity the medium should have in order to capture it. In addition the medium should be able to desorb the protein as the salt concentration decreases. Once proteins are captured too tightly to be eluted, organic solvent must be added to increase the elution power, which possibly causes the inactivation of proteins.

3.2. Mobile phase

Contrary to the IEXC, the initial buffer in HIC requires the presence of high concentration of salt ions, which preferentially take up the ordered water molecules from the protein surface and promote the hydrophobic interaction. The power is various among different ions. An ion that more increases the tension of water tend to more increase the strength of interaction between proteins and HIC media, although the internal nature is still not clear. Hofmeister series list the common ions according to the power to increase the water tension [10].

Anions: $HPO_4^{2-} > SO_4^{2-} > C_2H_3O_2^- > F^- > Cl^- > Br^- > I^- > ClO_4^- > SCN^-$

Cations: $N(CH_3)_4^+ > Cs^+ > Rb^+ > NH_4^+ > K^+ > Na^+ > Li^+ > Ca^{2+} > Mg^{2+}$

Molal surface tension of salts is listed as below.

$MgCl_2 > Na_2SO_4 > K_2SO_4 > (NH_4)_2SO_4 > MgSO_4 > Na_2HPO_4 > NaCl > LiCl > KSCN$

This series is not consistent for every protein, since except for the effect on water tension, the specific interaction between ions and proteins also appears to be another parameter on hydrophobic interaction. It seems that the hydrophobic interaction is more effected by anions that by cations. For example, the $MgCl_2$ is weaker than $(NH4)_2SO_4$ on the promotion of hydrophobic interaction.

In practice, $(NH_4)_2SO_4$ is one of the most used salt, 1~1.5 M of $(NH_4)_2SO_4$ solution could satisfied most protein separations. If could not obtain the ideal effect, altering concentration or changing other salt ions, such as Na_2SO_4 or NaCl, should be considered. The disadvantage of $(NH_4)_2SO_4$ is that the NH_4^+ tend to form ammonia gas under high OH^- concentration, so it should be used under pH < 8.0. As adding high concentration of salt into sample, some high hydrophobic proteins likely precipitate. Therefore ever remember to filter or centrifuge sample solution to remove particles after unstable proteins sufficiently aggregate.

Solution pH value also has complex effect on strength of hydrophobic interaction. The mechanism is not very clear. In general, an increase in pH weakens hydrophobic interaction [11], possible due to an increase of surface net charge. But a research of Hjerten et al. revealed that increase in pH, on the contrary, increased the retention of some protein [12].

The effect of temperature on hydrophobic interaction is also complex. An increase in temperature could promote the hydrophobic interaction for some proteins, but weaken it for some others. The effect still can not be predicted efficiently on theory.

3.3. Elution

Similar with IEXC, isocratic elution with constant solvent composition can not elute protein efficiently. Gradient decrease of ion strength is the mostly used method in elution process of HIC. By decrease of ion strength, proteins are desorbed in the order of increasing surface hydrophobicity.

As decrease of salt concentration, proteins again obtain ordered water molecules and are eluted in the order of increasing hydrophobicity. A linear or stepwise gradient decrease of salt concentration is employed in elution of protein in IHC. Similar to the strategies of IEXC, simple linear gradient elution presents even resolution to universal gradient range, which always used in the screening experiment or analytical separation, but takes more time. Stepwise gradient elution is preferred in large scale preparative separation. It is advantageous in time-saving and solution-saving and obtaining more concentrated product. But this strategy usually can not be performed until an appropriate elution condition is found out through preliminary works of linear gradient elution. A typical linear gradient elution spectrum is show in Figure 7.

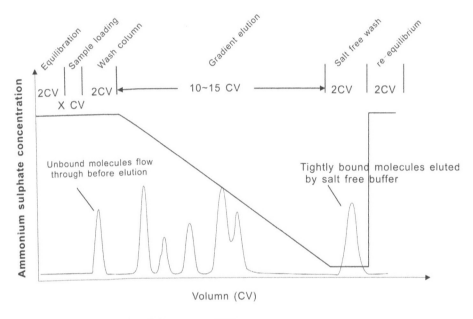

Figure 7. A typical linear gradient elution spectrum of HIC

Additionally, adding neutral nonpolar solution, such as detergents, to the elution buffer could promote the elution of higher hydrophobic protein, such as membrane proteins or aplipopro-teins. But nonpolar solution possibly causes irreversible inactivation, so should avoid to be used in IHC. If the target protein could not be eluted in salt free aqueous solution, changing of a lower hydrophobic medium should be considered. While, high concentration of organic solution could be used in column regeneration, by which tightly bound compounds will be washed away.

pH and temparature are two important factors on retention of proteins, but they are usually not used as variable parameters in elution since their effects are hardly controlled. So that, the pH and temperature condition should be consistent between patches in order to present a good reproducibility.

3.4. Features

HIC separates proteins based on different hydrophobicity of proteins. It combines the reversibility of hydrophobic interaction and the precision of column chromatography to yield excellent separation. With certain medium, HIC could capture almost all proteins at certain conditions and suit to capture, concentrate, or polish proteins.

The selectivity of HIC is orthogonal to that of IEXC and SEC, because it works base on hydrophobicity of proteins, a totally different property from the net surface charge used in IEXC and molecular size in SEC. So HIC is an orthogonal separation dimension when combining with IEXC or SEC. So using two of them in series will yields much better separation rather than using one.

4. Reversed-phase chromatography

Reversed-phase chromatography was named due to a reversed polarity between mobile phase and stationary phase compared with normal phase chromatography [13]. In normal phase chromatography, the mobile phase is organic solvent and stationary phase is hydrophilic resin. Reversely RPC uses hydrophobic adsorbents as stationary phase, which is the same with HIC in theory. However, in practice, the two methods have many differences. It is mainly due to the different degree of substitution of hydrophobic ligands on the medium surface. As shown in table 4, the density of ligand in RPC is an order of magnitude higher than that of HIC. It means that a protein molecule could bind more ligands when it is adsorbed. The huge forces could extract proteins from aqueous solution without help of neutral salt, so that the adsorbed proteins could not be eluted until using nonpolar solvents. Therefore, RPC is less used in preparation of activity proteins. However, the excellent resolution makes this technique to be the most important analytic chromatography. Liquid Chromatrography-Mass Spectrometry is an important extended application of the technique.

	RPC	HIC
Interaction	Hydrophobic interaction	Hydrophobic interaction
Ligand	C2~C8 alkyl or aryl	C4~C18 alkyl
Substitution degree	10–50 mmoles/ ml gel	several hundred mmoles/ml gel
Capture condition	Salt free solution	High salt solution
Elution	Increase nonpolarity	Decrease ion strength
Application	Protein analysis Preparative separation of poly peptide or oligonucleotide	Preparative separation of protein

Table 4. Comparison between RPC and HIC

4.1. Stationary phase

Similar with HIC, the media of RPC is composed of inert base matrix and hydrophobic ligands on surface.

The base matrix for reversed phase media is generally composed of silica or a synthetic organic polymer such as polystyrene. Silica was the first material used as base matrix for RPC, which has an excellent mechanical strength and chemical stability under acid condition. However the disadvantages of silica base matrix is its chemical instability in aqueous solutions at high pH. Silica matrix could be dissolved at high pH, so it is not recommended for prolonged exposure above pH7.5. Additionally, due to incomplete substitution or long term usage, some underivatised silanol groups are exposed to mobile phase, which will be negatively charge at high pH value, and cause ionic interaction with proteins. The mixed chromatography always causes decreased resolution with significant broadening and tailing of peaks. Therefore, RPC using silica matrix is often performed at low pH values (<3).

The loading capacity and resolution are determined by size of resin, in general, smaller resin give the higher resolution but lower loading capacity. The resin with 3~5 μm in diameter is preferable for analytic separation. Due to small size, it is hard to be packed well. So it is often offered in the form of prepacking columns. With increasing of diameter, the loading capacity increase, but resolution decrease simultaneously. Generally media with 15 μm or larger diameter are used in preparative separation.

The porous structure is employed to increase the loading capacity of PRC media. In general the pore size is 10~30 nm. Media with pore sizes of 10 nm are used predominately for small peptides or molecules. Media with pore sizes of 30 nm or greater are used in purification of large peptide or proteins.

Ligands used in RPC are linear alkyl with different length of carbon chain, which is the main factor on selectivity of media. In general, a medium with longer chain ligands gives stronger hydrophobicity. Oligonucleotide and organic moleculars, having less hydrophobicity, needs more hydrophobic media to supply sufficient adsorbability, such as C18 media. On the contrary, large peptides or proteins generally have more hydrophobic sites and need less hydrophobic adsorbents, such as C4 or C8. Selectivity and loading capacity are also influenced by the substitution degree. For large peptides or protein, the effect of increase in substitution degree is equal to increase in length of carbon-chain.

4.2. Mobile phase

4.2.1. Organic solvent

Typically, sample was loaded onto the column in aqueous solution and eluted by decreasing solution polarity. The elution power increases as polarity decreases. Although a large part of organic solvents have enough elution power, only a few of them could be used in RPC because of the requirement on viscosity and ultraviolet (UV) transparence. High solution viscosity influences the diffusion of solutes between mobile and stationary phases, therefore high viscous solvent reduces resolution. UV absorption of solvent will disturb the detection of solute

UV absorption. Acetonitrile and methanol are two most widely used organic modifiers due to their moderate viscosity and perfect UV transparent. Although isopropanol and normal propanol have higher elution power, they are only used to clean and regenerate column because of their high viscosity.

It should be noted, all solvent used in RPC should be HPLC grade to minimize the damage of impurities to resin or samples.

4.2.2. pH

pH value could influence protein hydrophobicity by possibly changing the charge property of proteins [14]. In practice, two proteins with the same retention time are likely separated by just changing the solution pH value, and *vise versa*. At present, there is not effective method to predict the effect, trying different pH value is the only way to optimize the resolution.

However, as described above, media base on silica matrix are not suit to work at high pH value because of uncovered silanol groups. So silica-based RPC should works at low pH value, in general between 2 to 3. Strong acids, such as trifluoroacetic acid (TFA) or ortho-phosphoric acid are typically used to just the pH.

4.2.3. Ion-pairing agent

The retention time of solutes, such as proteins, peptides, or nucleotides can be modified by adding ion pairing agents to solution [15]. An ion-pairing agent could ionize and release positive or negative ions, which will bind to the sample molecules by ionic interactions and results in the modification of hydrophobicity. For example, at a very acid condition most proteins are positively charged. The negative ion pairing agent will bind to positive charge group. The effect of neutralization always increases the hydrophobicity of proteins. TFA is not only used in pH control but is the most commonly used negative ion pairing agent. Additionally, triethylamine is used as positive ion pairing agent in neutral and alkaline condition.

4.3. Elution

A simple linear gradient elution is often used in RPC. The eluent is a mixture of buffer A and buffer B by a mix pump. The buffer A generally is the start buffer, in which 0.1~0.5% TFA is added to control pH and functions as an ion pairing agent. The Buffer B typically is 0.1~0.5% TFA in pure organic solvent, such as acetonitrile or methanol. A gradient increase of buffer B from 0% to 90% or more in 30~60 min is often used.

4.4. Application

The application of RPC on protein separation is mainly focus on the analytic separation and purity check. Because, on one hand, RPC has the highest resolution compared with the other relative techniques, on the other hand, the harsh binding and desorption condition in RPC usually leads to protein denaturation and not suit to preparative separation. A good reproducibility on retention time and low limit of detection make it be the most favored method in protein purity check. Additionally, RPC is the only one chromatography that can be used in association with mass spectrometry analysis, since the high resolution of RPC is the only one chromatography can separate a complex sample, such as serum, into single components and immediately analyzed by mass spectrometry.

5. Size exclusion chromatography gel filteration chromatography

Size exclusion chromatography (SEC), or termed gel filtration chromatography, separates protein according to the difference on molecular size [16]. Different to those chromatography techniques based on adsorption, molecules do not bind to the surface of media in SEC, but are retarded by the porous structure of media. As shown in Figure 8, media of SEC are composed of porous material. However the pore size is much smaller than the pore size of the matrix used in adsorption chromatography and not uniform. The pore size of adsorption chroma-tography is big enough to allow entries of all molecules without selectivity. Comparatively, the pore sizes of SEC are smaller and selectively allow molecules with appropriate size enter and exclude the bigger molecules outside. Smaller molecules run longer and more winding paths in media rather than running straight paths outside the media as larger molecules do. So that smaller molecules are more retarded than larger ones.

5.1. Stationary phase

Resolution of SEC is influenced by many parameters of stationary phase, including, column volume, particle size, pore size distribution [17].

The matrix of SEC are often composed of polymers by cross-linking to form a three-dimen-sional network. The matrix is manufactured in small spherical particles. On the surface and the inside of the particles, small channels and pores are formed with different sizes by controlling different degree of cross-linking. The selectivity of a medium depends on the distribution of pore sizes and can be described by a selectivity curve (Figure 9). For example, the medium superdex 200 (by GE company) has a linear selectivity range of $1\times10^4\sim6\times10^5$, that means solutes having molecular mass (Mw) in this range could be differentially retarded. The molecules larger than the upper limit are completely excluded from the inside space of the medium because no pores are big enough to allow them enter. At this time, the distribution coefficient (Kd) reaches to 0. On the contrary, those molecules smaller than the lower limit are free to enter any channel, therefore they are maximally retarded without selectivity and has a Kd=1. Those solutes with Mw between the two extremes could enter channels with different degree, Kd is between 0 and 1, are retarded differentially.

The media with narrow linear range often employed in group separations, by which solutes are simply separated into two groups. A typical application is protein desalting by a G25 column (Figure 9). On the contrary, the media with wide linear range usually used to separate similar components (Figure 9), such as using superdex 200 to separate IgG (Mw=1.5×10^5) and albumin (Mw=6.6×10^4).

Figure 8. In SEC large molecules run though the space between media with a shorter pathway, while the smaller molecules run through the channels inside the medium with a longer pathway.

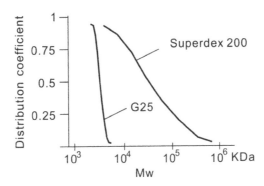

Figure 9. Selectivity curves of Superdex 200 and G25 media.

The height of packing bed affects both resolution and the separation time. Larger bed height often gives a better resolution with same sample volume, but takes more time to run a separation (Figure 10C).

The size of particle also is a parameter affecting resolution and the separation time. Smaller resin particles supply more efficient mass transfer between mobile and stationary phase, therefore present higher resolution. But simultaneously smaller particles increase the flow resistance and generally cause prolonged separation time.

5.2. Mobile phase

An unparalleled advantage of SEC in all chromatography is the wide compatibility to various solutions. Because SEC separates proteins depends on molecular size rather than interactions between solutes and media, so pH value and polarity of mobile phase generally have slight influence the retention of compounds.

Since SEC has no concentration effect on elutes, so volumes of elution peak of each components are proportional to the sample volume. Increased sample volume will decrease the resolution (Figure 10B).

High viscosity in mobile phase has a certain effect on resolution by influence on the mass transfer between the mobile and the stationary phases, so that will cause broadening and tailing peaks (Figure 10D).

It is should be noticed that the ionic interaction between proteins and the resin possibly takes place at a low ionic strength, so generally 0.15 M NaCl is added to avoid it.

5.3. Elution

SEC has no a definite elution step, since molecules are not adsorbed by media. After sample is loaded, a buffer usually same to the initial buffer is pumped with two column volumes until all solutes are eluted.

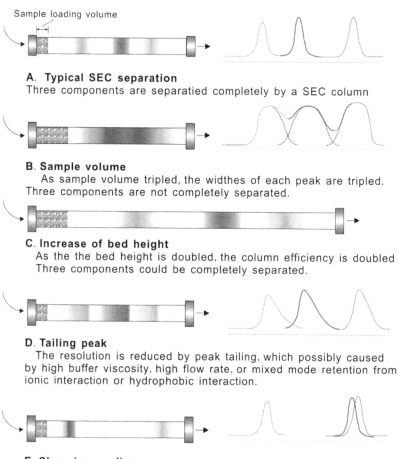

A. Typical SEC separation
Three components are separatied completely by a SEC column

B. Sample volume
As sample volume tripled, the widthes of each peak are tripled. Three components are not completely separated.

C. Increase of bed height
As the the bed height is doubled, the column efficiency is doubled Three components could be completely separated.

D. Tailing peak
The resolution is reduced by peak tailing, which possibly caused by high buffer viscosity, high flow rate, or mixed mode retention from ionic interaction or hydrophobic interaction.

E. Changing medium
After a target component (the blue one) is determined, a medium suit to large molecules separation is used, which increases the resolution between high Mw components, but sacrifices the resolution of components at low Mw range

Figure 10. The factors affecting resolution of SEC.

5.4. Application

SEC has the most mild separation condition, since in the whole process the composition of mobile phase needs no change. This is a good property for separating proteins that are unstable to alterations of pH value, ionic strength or polarity. SEC is often used in polish step after a sample has been crudely separated by other chromatography, especially in separation of the monomer and polymers. Since monomer and polymers usually could not be separated by IEXC and HIC due to the similarities in charge and hydrophobic property. But fortunately SEC can well separate them by different molecular size.

5.4.1. Purification of recombinant human Midkine by SP column and SEC column

A recombinant human Midkine (Mw=14 kDa) was expressed by an *E.coli* BL21 strain as inclusion body form. The inclusion body was denatured by 6 M guanidinium chloride and renatured through 10-fold dilution in renature buffer. The renatured protein was separated by IEXC and SEC (figure 11). Since the incorrect formation of intermolecular disulfide bond, a fraction of the rhMK molecules formed different polymers, which could not be separated from monomers by IEXC and were eluted as a mixture (Figure 11A). To separate bioactive monomers, a Sephadex G-75 column, which owns a fractionation range of 3000~80,000 dalton, was used to separate monomers from polymers. Non-reduced SDS PAGE demonstrated the purity of monomers reached 95% in the target peak.

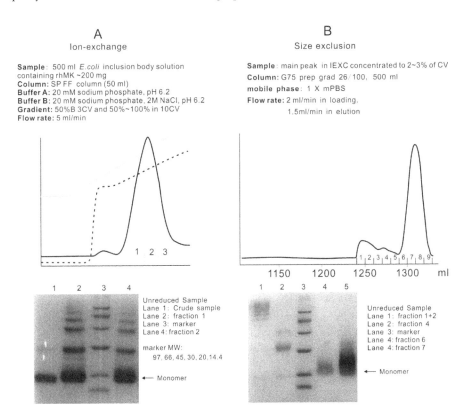

Figure 11. Purification of E. coli rhMK by IEXC and SEC. (result of Shixiang Jia, Ping Tu et al. General regeneratives (shanghai) limited, Shanghai, PR China)

6. Affinity chromatography

Affinity chromatography (AC) extensively refers to a series of techniques that separate proteins on the basis of a reversible interaction between proteins and their specific ligands coupled to a chromatography matrix [18]. The affinity interactions derive from a wide range of biorecognition, briefly including interactions between (1) enzymes and substrate analogues, inhibitors, cofactors [19], (2) antibodies and antigens [20], (3) membrane receptors and ligands [21], (4) nucleic acid and complementary sequence, histones, or nucleic acid polymerase, nucleic acid binding proteins, (5) biological small molecules and their receptors or carrier proteins [22], (6) metal ions and proteins having polyhistidine sequence.

Affinity interactions are always a result of a combination of different types of interactions, including electrostatic interactions, hydrophobic interactions, van der Vaals' forces, or hydrogen bonding. The interactions of high specificity always supply extremely high selectivity, by which a target protein could easily be separated in one step with thousands fold of increase in purity and high recovery.

6.1. Media

Development of an AC media is much more complex than that other chromatography. It needs not only a specific ligand, but also complex coupling process to couple the ligand to the matrix without reducing its binding activity significantly. Therefore more and more ready-to-use matrices, which already have active ligands coupled to, were developed commercially to satisfy different separation. If no suitable ligand is available, it can be considered to develop a specific affinity medium or use alternative purification techniques.

6.1.1. Base matrix

The mostly used material is agarose or cross-linked agarose. The hydroxyl groups on the sugar resides are easily derivatized for covalent attachment of a ligand or spacer arms and the porous structure also supplies ideal flow rate and high capacity.

6.1.2. Spacer arms

The binding site of a target protein often locates deep within the molecule. Due to steric interference, a small ligand directly coupled to the matrix always shows a lower affinity with the target protein than in their free state. To overcome this situation, spacer arms, typically linear molecules with different chain length, are used to bridge ligands and matrix. In general a spacer arm is necessary in coupling ligands Mw <1000, and not need for larger ligands (Figure 12). An ideal spacer arms should have active groups at two ends by which it can be covalently coupled with matrix and ligand respectively. After coupling with matrix and ligand, the arms should be chemically stable to avoid reaction with other solutes and be hydrophilic to avoid the hydrophobic interaction with proteins.

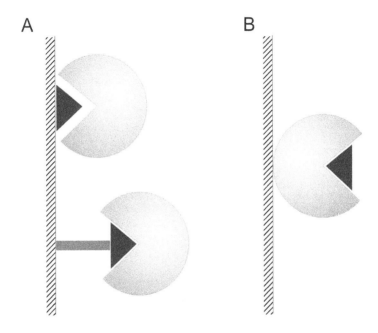

Figure 12. The influence of spacer arms on small or large ligands. A spacer arm is often necessary for coupling small ligands, which ensure a efficient binding between ligands and target proteins (A), but not necessary for large ligands (B).

The atom number of commonly used space arms varies from 4 to 12. They often coupled with
agarose matrix by stable ether links at one end and with ligand by other chemical bonds at the
opposite end.

6.1.3. Ligand coupling

A coupling procedure of ligand is generally composed of three steps. First a group on matrix
or spacer arm is activated by an activating agent. And then the activated group reacts with a
functional group on ligand molecules. Finally, residual unreacted groups are blocked by
blocking agent [23]. A matrix can be coupled with a ligand by a chemical group on itself or by
groups on spacer arms. A variety of spacer arms are available to couple with to functional
groups on ligands such as amino, hydroxyl, carboxyl, thiol groups (Figure 13).

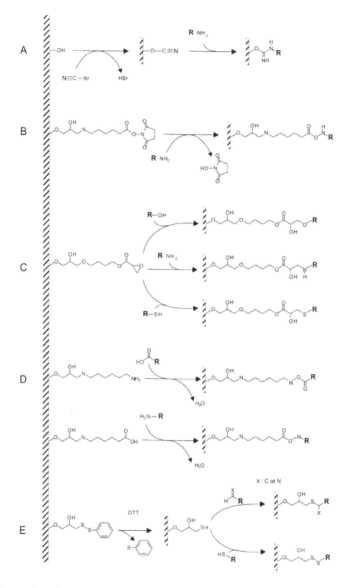

Figure 13. Commonly used spacer arms and immobilization procedures of ligands. (A) Ligands are directly coupled with matrix by reaction between Cyanogen bromide activated hydroxyl on matrix and amino group on ligand. (B) Ligands are coupled with spacer arms by reaction between N-hydroxysuccinimide activated carboxyl and amino group on ligand. (C) Lgands couple with spacer arms by reaction with epoxy group. (D) Coupling through condensation between a free amino and a free carboxyl group. (E) Coupling through bisulfide bond or additive reaction between silanol and double bond in ligand, such as N=N or C=N.

6.1.4. Steric interference

For a small ligand, it should be paid attentions to the influences of steric interference even if a spacer arm has been used. For small ligands the amount of each functional group is rare. even just one. A bad choice that makes a wrong spatial orientation in coupling will likely cause a serious decrease in binding capacity or even complete failure. On the contrary, large ligands have several equivalent groups through which coupling takes place, so that a large proportion couplings leave sufficient space for binding with target molecules (Figure 14). Therefore in coupling a small ligand, it is important to choose a suitable functional group without introducing significant steric interference. The information of structure can be obtained from databases of X-ray crystal diffraction or NMR, or prediction by computational biology.

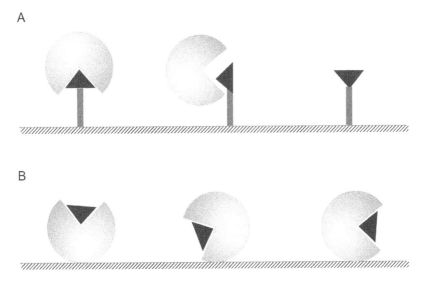

Figure 14. The influences of steric interference to small and large ligands. (A) For a small ligand, an inappropriate coupling orientation likely results in steric interference and inefficient adsorption. (B) This situation is less happened on large ligands.

6.2. Binding and elution

An ideal binding buffer should be optimized to ensure efficient interaction between target molecules and ligands and minimize the nonspecific interaction at same time. Since the ligand-protein interaction is a result of combination of electrostatic attraction, hydrophobic interaction and hydrogen bonds, the binding conditions can be optimized on these aspects.

Adsorbed proteins could be eluted by modification of pH value, ionic strength, or polarity. pH value could be decrease to pH 2~3 to reduce the charge property of interaction surface between proteins. For example, immunoglubin could be adsorbed by a protein A column and eluted by a glycine buffer with pH 3.0. But the eluted sample should be neutralized as soon as possible to avoid being destroyed in extreme circumstance.

The ionic interaction also can be weakened by adding neutral salt, for example 1M NaCl is frequently used in practice.

A specific elution can be performed by adding competitors of either ligands or target proteins in elution buffer. An ideal competitor should have a moderate dissociation coefficient to the ligand or the target molecule, so that the competitor can elute target with high concentration but can be easily removed from column by wash or isolated from target protein by dialysis. Two classic applications is affinity chromatography of Glutathione S-transferase (GST) and polyhistdine [24] (Figure 15).

In binding process, flow rate should be control at a relative low degree to ensure an effective binding capacity.

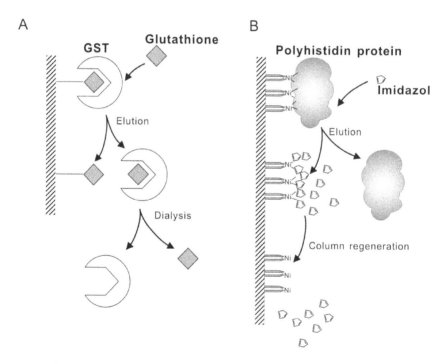

Figure 15. Different elution mechanisms in GST affinity chromatography and metal chelate interaction chromatography (A) In GST purification, GST is captured by a medium with immobilized glutathione, and then dissociated by adding excess reduced glutathione. The excess glutathione is eluted together with target protein and removed by dialysis. (B) In nickel ion chelate interaction chromatography, protein with polyhistidine sequence is adsorbed by a medium with immobilized ionized nickel through a chelation between nickel ion and imidazolyl on polyhistidine sequence. The protein is eluted by adding high concentration of imidazol, a competitor of the imidazolyl on the protein. Finally, the small competitor is washed away from column by binding buffer.

6.3. Tag purification strategy

AC separates protein typically on the basis of interactions between ligands and local domains of target proteins. The interactions are not interfered by other domains in most case. Therefore, the tag purification strategy was invented to rapidly separate recombinant protein by fusion expression and co-separation [25].

First the target protein is expressed with a tag protein in fusion form. Then the target protein is purified using an affinity column that is specific to tag protein. After that if the tag needs to be removed, a restrictive protease is used to hydrolyze the fusion protein and the freed tag is finally be separated from target protein by running the same column once again.

The ideal tag protein should (1) have economical affinity chromatography media for convenient separation, (2) be very stable in bioactivity, and (3) have a good expressing property that is helpful to increase the expression of target protein. Commonly used tags are GST tag, FLAG tag, S tag, Strep tag, His tag, and so on.

6.4. Application

Affinity chromatography is a rapid and efficient chromatography technique. The high specific biorecognition give the technique an extremely high selectivity, by which a protein or a group of proteins could be separated from a crude sample in one step and reaches to a satisfying purify. However the excellent performance is based on the complex productive technology. Development of each noval medium needs a plenty of trials on finding suitable ligand and coupling the ligand on matrix properly. It is worth time and effort to develop a new specific affinity medium for high scale protein production, otherwise, the alternative method such as tag purfication or other chromatography should be a better choice for small scale preparation in expiremental research.

7. Summary

This chapter introduces principles and applications of several basic chromatography techniques. Different techniques separate proteins depending on different properties including net surface charge, hydrophobicity, molecular size, and affinity interaction. Affinity chromatography has the highest selectivity and can purify target proteins in one step to > 95% purify. But due to the difficulties on obtaining and immobilization of suitable ligand, this chromatography technique is not used as widely as other ones. HIC and RPC are both based upon hydrophobic interaction. PRC is widely used in analytic separation because of its high resolution, but less used on preparative separation of proteins since the high nonpolarity of the eluent likely causes irreversible inactivation of proteins. IEXC, HIC and SEC separate proteins in mild conditions and are suitable for large scale separation of active proteins. However, their resolutions are comparatively lower and hard to purify a protein from complex components by a single technique. An ideal purification could be achieved by combined application of several techniques.

Author details

Jingjing Li[1], Wei Han[1] and Yan Yu[2]

1 Laboratory of Regeneromics, School of Pharmacology, Shanghai Jiao Tong University, Shanghai, China

2 School of Agriculture and Biology, Shanghai Jiao Tong University, Shanghai, China

References

[1] Zechmeister, L. Early history of chromatography, *Nature*, (1951).

[2] Dorsey, J. G, Foley, J. P, & Cooper, W. T. Liquid chromatography: theory and methodology, *Anal Chem*, (1990). R-356R.

[3] Hough, L, Jones, J. K, & Wadman, W. H. Application of paper partition chromatography to the separation of the sugars and their methylated derivatives on a column of powdered cellulose, *Nature*, (1948).

[4] Thijssen, H. A. Gas-liquid chromatography. A contribution to the theory of separation in open hole tubes, *J Chromatogr*, (1963).

[5] Dorsey, J. G, Cooper, W. T, & Wheeler, J. F. Liquid chromatography: theory and methodology, *Anal Chem*, (1994). R-546R.

[6] Cashman, P. J, & Thornton, J. I. High speed liquid adsorption chromatography in criminalistics. I. Theory and practice, *J Forensic Sci Soc*, (1971).

[7] Woods, M. C, & Simpson, M. E. Purification of sheep pituitary follicle-stimulating hormone (FSH) by ion exchange chromatography on diethylaminoethyl (DEAE)-cellulose, *Endocrinology*, (1960).

[8] Melander, W. R, Corradini, D, & Horvath, C. Salt-mediated retention of proteins in hydrophobic-interaction chromatography. Application of solvophobic theory, *J Chromatogr*, (1984).

[9] Biswas, K. M, Devido, D. R, & Dorsey, J. G. Evaluation of methods for measuring amino acid hydrophobicities and interactions, *J Chromatogr A*, (2003).

[10] Zhang, Y, & Cremer, P. S. Interactions between macromolecules and ions: The Hofmeister series, *Curr Opin Chem Biol*, (2006).

[11] Porath, J, Sundberg, L, & Fornstedt, N. Salting-out in amphiphilic gels as a new approach to hydrophobic adsorption, *Nature*, (1973).

[12] Parente, E. S, & Wetlaufer, D. B. Relationship between isocratic and gradient reten-
     tion times in the high-performance ion-exchange chromatography of proteins. Theo-
     ry and experiment, *J Chromatogr*, (1986).

[13] Molnar, I, & Horvath, C. Reverse-phase chromatography of polar biological substan-
     ces: separation of catechol compounds by high-performance liquid chromatography,
     Clin Chem, (1976).

[14] Sottrup-jensen, L. A low-pH reverse-phase high-performance liquid chromatography
     system for analysis of the phenylthiohydantoins of S-carboxymethylcysteine and S-
     carboxyamidomethylcysteine, *Anal Biochem*, (1995).

[15] White, E. R, & Zarembo, J. E. Reverse phase high speed liquid chromatography of an-
     tibiotics. III. Use of ultra high performance columns and ion-pairing techniques, *J An-
     tibiot (Tokyo)*, (1981).

[16] Kostanski, L. K, Keller, D. M, & Hamielec, A. E. Size-exclusion chromatography-a re-
     view of calibration methodologies, *J Biochem Biophys Methods*, (2004).

[17] Paul-dauphin, S, Karaca, F, & Morgan, T. J. Probing Size Exclusion Mechanisms of
     Complex Hydrocarbon Mixtures: The Effect of Altering Eluent Compositions, *Energy
     Fuels*, (2007).

[18] Chaiken, I. M. Analytical affinity chromatography in studies of molecular recogni-
     tion in biology: a review, *J Chromatogr*, (1986).

[19] Caldes, T, Fatania, H. R, & Dalziel, K. Purification of malic enzyme from bovine heart
     mitochondria by affinity chromatography, *Anal Biochem*, (1979).

[20] Santen, R. J, Collette, J, & Franchimont, P. Partial purification of carcinoembryonic-
     reactive antigen from breast neoplasms using lectin and antibody affinity chromatog-
     raphy, *Cancer Res*, (1980).

[21] Bluestein, B. I, & Vaitukaitis, J. L. Affinity chromatography purification of solubilized
     FSH testicular membrane receptor, *Biol Reprod*, (1981).

[22] Yamada, S, Itaya, H, & Nakazawa, O. Purification of rat intestinal receptor for intrin-
     sic factor-vitamin B-12 complex by affinity chromatography, *Biochim Biophys Acta*,
     (1977).

[23] Healthcare, G. Affinity Chromatography Principles and Methods, (2007).

[24] Scheich, C, Sievert, V, & Bussow, K. An automated method for high-throughput pro-
     tein purification applied to a comparison of His-tag and GST-tag affinity chromatog-
     raphy, *BMC Biotechnol*, (2003).

[25] Li, Y, Franklin, S, & Zhang, M. J. Highly efficient purification of protein complexes
     from mammalian cells using a novel streptavidin-binding peptide and hexahistidine
     tandem tag system: application to Bruton's tyrosine kinase, *Protein Sci*, (2011).

Chapter 2

Protein-Protein and Protein-Ligand Docking

Alejandra Hernández-Santoyo,
Aldo Yair Tenorio-Barajas, Victor Altuzar,
Héctor Vivanco-Cid and Claudia Mendoza-Barrera

Additional information is available at the end of the chapter

1. Introduction

Molecular interactions including protein-protein, enzyme-substrate, protein-nucleic acid, drug-protein, and drug-nucleic acid play important roles in many essential biological proc-esses, such as signal transduction, transport, cell regulation, gene expression control, enzyme inhibition, antibody–antigen recognition, and even the assembly of multi-domain proteins. These interactions very often lead to the formation of stable protein–protein or protein-ligand complexes that are essential to perform their biological functions. The tertiary structure of proteins is necessary to understand the binding mode and affinity between interacting molecules. However, it is often difficult and expensive to obtain complex structures by experimental methods, such as X-ray crystallography or NMR. Thus, docking computation is considered an important approach for understanding the protein-protein or protein-ligand interactions [1-3]. As the number of three-dimensional protein structures determined by experimental techniques grows —structure databases such as Protein Data.

Bank (PDB) and Worldwide Protein Data Bank (wwPDB) have over 88000 protein structures, many of which play vital roles in critical metabolic pathways that may be regarded as potential therapeutic targets — and specific databases containing structures of binary complexes become available, together with information about their binding affinities, such as in PDBBIND [4], PLD [5], AffinDB [6] and BindDB [7], molecular docking procedures improve, getting more importance than ever [8].

Molecular docking is a widely used computer simulation procedure to predict the conforma-tion of a receptor-ligand complex, where the receptor is usually a protein or a nucleic acid molecule and the ligand is either a small molecule or another protein (Figure 1).

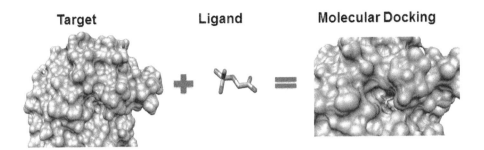

Figure 1. Elements in molecular docking.

The accurate prediction of the binding modes between the ligand and protein is of fundamental importance in modern structure-based drug design. The most important application of docking software is the virtual screening, in which the most interesting and promising molecules are selected from an existing database for further research. This places demands on the used computational method: it must be fast and reliable. Another application is the research of molecular complexes.

Since the pioneering work of Kuntz *et al.* [9] during the early 1980s, significant progress has been made in docking research to improve the computational speed and accuracy. Over the last years several important steps beyond this point have been given. Handling efficiently the flexibility of the protein receptor is currently considered one of the major challenges in the field of docking. The binding-site location and binding orientation can be greatly influenced by protein flexibility. In fact, X-ray structure determination of protein–ligand complexes frequently reveals ligands with a buried surface area in the range of 70–100%, which can only be achieved as a consequence of protein flexibility [3]. There are many interesting docking suites and algorithms that have shown significant progress in predicting near-native binding poses by making use of biophysical and biochemical information combination with bioinformatics.

2. Theory

Modeling the interaction of two molecules is a complex problem. Many forces are involved in the intermolecular association, including hydrophobic, van der Waals, or stacking interactions between aromatic amino acids, hydrogen bonding, and electrostatic forces. Modeling the intermolecular interactions in a ligand-protein complex is difficult since there are many degrees of freedom as well as insufficient knowledge of the effect of solvent on the binding association. The process of docking a ligand to a binding site tries to mimic the natural course of interaction of the ligand and its receptor via the lowest energy pathway [3]. There are simple methods for docking rigid ligands with rigid receptors and flexible ligands with rigid receptors, but general methods of docking considering conformationally flexible ligands and receptors are problematic. Docking protocols can be described as a combination of a search algorithm, and the scoring functions (Figure 2).

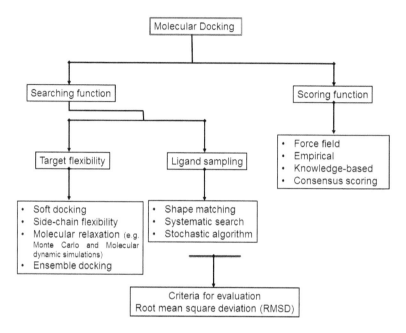

Figure 2. Methods used for protein-ligand docking.

The search algorithm should create an optimum number of configurations that include the experimentally determined binding modes. Although a rigorous searching algorithm would go through all possible binding modes between the two molecules, this search would be impractical due to the size of the search space and amount of time it might take to complete. As a consequence, only a small amount of the total conformational space can be sampled, so a balance must be reached between the computational expense and the amount of the search space examined. Some common searching algorithms include molecular dynamics, Monte Carlo methods, genetic algorithms, fragment-based, point complementary and distance geometry methods, Tabu, and systematic searches. On the other hand, scoring function consists of a number of mathematical methods used to predict the strength of the non-covalent interaction called the binding affinity. In all the computational methodologies, one important problem is the development of an energy scoring function that can rapidly and accurately describe the interaction between the protein and ligand. Several reviews on scoring are available in the literature [10-12].

There are three important applications of scoring functions in molecular docking. First is to determine the binding mode and site of a ligand on a protein. The second application is to predict the absolute binding affinity between protein and ligand. This is particularly important in lead optimization. The third application, and perhaps the most important one in structure-based drug design, is to identify the potential drug hits/leads for a given protein target by searching a large ligand database. Over the course of the last years, different scoring functions have been developed that exhibit different accuracies and computational efficiencies. Some of these commonly-used functions are: force-field, empirical, knowledge-based and consensus scoring.

The protein-ligand docking procedure can be typically divided into two parts: rigid body docking and flexible docking [9].

1. *Rigid Docking.* This approximation treats both the ligand and the receptor as rigid and explores only six degrees of translational and rotational freedom, hence excluding any kind of flexibility. Most of the docking suites employ rigid body docking procedure as a first step.

2. *Flexible Docking.* A more common approach is to model the ligand flexibility while assuming having a rigid protein receptor, considering thereby only the conformational space of the ligand. Ideally, however, protein flexibility should also be taken into account, and some approaches in this regard have been developed. There are three general categories of algorithms to treat ligand flexibility: systematic methods, random or stochastic methods, and simulation methods [3]. Due to the large size of proteins and their multiple degrees of freedom, their flexibility may be the most challenging issue in molecular docking. The methods to address the flexibility of proteins can be grouped into: soft docking, side-chain flexibility, molecular relaxation and protein ensemble docking. They were described by Huang *et al* [1].

3. Experimental docking procedures

There are a number of excellent reviews of molecular docking methods and a large number of publications comparing the performance of a variety of molecular docking tools [1-3], [13]. Following, we will describe the four-step procedure adopted in this study to perform the molecular docking.

3.1. Target selection

Ideally, the target structure should be determined experimentally by either X-ray crystallography or nuclear magnetic resonance, which can be downloaded from PDB; however, docking has been performed successfully in comparison to homology models or threading. The model should have good quality. It can be tested using validation software such as Molprobity [14]. After selecting the model, it must be prepared by removing the water molecules from the cavity, stabilizing charges, filling the missing residues, and generating the side chains, all according to the available parameters. The receptor should be at this point biologically active and in the stable state.

3.2. Ligand selection and preparation

The type of ligands chosen for docking will depend on the goal. It can be obtained from various databases, e.g. ZINC or/and PubChem, or it can be sketched by means of Chemsketch tool [8]. Often it is necessary to apply filters to reduce the number of molecules to be docked. Examples include the net charge, molecular weight, polar surface area, solubility, commercial availability, similarity thresholds, pharmacophores, synthetic accessibility, and absorption, distribution, metabolism, excretion, and toxicology properties. Many times the researchers design their own molecules such as those generated by us in the example that will be described in this work in the section 5.

3.3. Docking

This is the last step, where the ligand is docked onto the receptor and the interactions are checked. The scoring function generates a score depending on the best selected ligand.

3.4. Evaluating docking results

The success of docking algorithms in predicting a ligand binding pose is normally measured in terms of the root-mean-square deviation (RMSD) between the experimentally-observed heavy-atom positions of the ligands and the one(s) predicted by the algorithm. The flexibility of the system is a major challenge in the search for the correct pose. The number of degrees of freedom included in the conformational search is a central aspect that determines the searching efficiency [3]. A good performance is usually considered when the RMSD is less than 2Å.

3.5. Docking software description

There are many algorithms available to assess and rationalize ligand-protein or protein-protein interactions, and their number is constantly increasing. Speed and accuracy are key features for obtaining successful results in docking approaches. Several algorithms share common methodologies with novel extensions focused on obtaining a fast method with accuracy as high as possible. The most common docking programs include AutoDock [15], DOCK [9], FlexX [16], GOLD [17], ICM [18], ADAM [19], DARWIN [20], DIVALI [21], and DockVision [22].

4. Application of molecular docking to a particular case — Biopolymers docked to dengue virus E protein

In the last decades, the incidence of Dengue disease has dramatically increased around the world. About 2.5 billion persons (two fifth of the world population) are exposed to the risk of contracting the disease. Every year, dengue virus (DENV) infects more than 50 million people, with approximately 22 000 fatal cases [23]. The disease is endemic in more than 100 countries of Africa, America, Oriental Mediterranean, Southeast Asia, and the Western Pacific Ocean with the last two regions being the most affected by the disease. Before 1970 only nine countries suffered from the Hemorrhagic Dengue (HD) epidemics, number that in 1995 was multiplied for more than four. There are four antigenically distinct, but closely related, serotypes of dengue virus (DENV), which is a Flavivirus member of the family Flaviviridae [24]. Each serotype has genotypes, which are virulent at several levels; nevertheless, the factors of virulence are not totally established [25]. A better understanding of the mechanisms and the molecules involved in the key steps of the DENV transmission cycle may lead to the identification of new anti-dengue targets [26]. In fact, the presence of two or more serotypes in the same geographical region implies a growing risk to population of contracting Hemorrhagic Dengue or Dengue Shock Syndrome (SSD) due to a phenomenon known as the Antibody – Dependent Enhancement (ADE). As a result, the diagnosis and treatment of dengue disease has become a world-wide global problem to deal with. The mature DENV virion contains three structural proteins: capsid protein (C), membrane protein (M), and envelope protein (E). In particular, the DENV E glycoprotein (51-60 kDa~ 495 aa), found on the viral surface, is important in the initial attachment of the viral particle to the host cell, as it contains two N-linked glycosylation sites at Asn-67 and Asn-153. While the glycosylation site at position 153 is conserved in most flaviviruses, the site at position 67 is thought to be unique for dengue virus. N-linked oligosaccharide side chains on flavivirus E proteins have been associated with viral morphogenesis, infectivity, and tropism [27, 28]. In addition, E protein is closely associated with the lipid envelope containing a cellular receptor-binding site (s) and a fusion peptide [29]. It can be found in a form of a homodimer on the surface of the mature virion, and inside the cell, it creates a prM-E heterodimer together with the prM protein. E protein is the principal component of the virion surface, containing the antigenic determinants (epitopes) responsible for the neutralization of the virus and the hemagglutination of erythrocytes, inducing thereby an immunological response in the infected host [29]. On native virions, the elongated three-domain E molecule is positioned tangentially to the virus envelope in a head-to-tail homodimeric conformation. Upon penetration of the virion into the target cell endosome, E dimers are converted into stable target-cell membrane-inserted homotrimers that reorient themselves vertically to promote virus-cell fusion at low pH [30]. Furthermore, there is a great deal of evidence that E protein contains the majority of molecular markers for pathogenicity. Comparing the nucleotide sequence of the E protein gene in different flaviviruses has demonstrated a perfect conservation of 12 cysteine residues, which form six disulfide bridges. The structural model for the E protein was refined by Mandl and co-workers [31], who correlated the structural properties of different epitopes with disulfide bonds [32].

4.1. Biopolymers as potential adjuvants carriers

The aim of this work is to study the docking of monomers of polyvinylpyrrolidone (PVP), chitosan (CS), and chitosan-tripropylphosphate-chitosan (CS/TPP/CS) with E protein of dengue virus in order to use them as potential adjuvant carriers. Given their structure, these polymers have specific molecular anchor sites that are expected to be exploited to induce antigenic specificity to the conserved regions of dengue virus. Several authors report that the E protein produces immunity and confers protection against infection in mice with low levels of neutralizing antibodies [33-35]. Because of the dual role of its receptors as well as the cell entry through membrane fusion, the E protein, apart from being the most exposed protein, is the main target against which the neutralizing antibodies are produced to inhibit its functions.

At present, the biggest challenge in developing an efficient dengue vaccine is to achieve a life-long protective immune response to all 4 serotypes (DEN1-4) simultaneously. Although several vaccines are currently being developed, so far only a chimeric dengue vaccine for live attenuated yellow fever (YF) has reached stage 3 in clinical trials. The candidate vaccines can be divided into the following types: (a) live attenuated, (b) DEN-DEN and DEN-YF live chimeric virus, (c) inactivated whole virus, (d) live recombinant, (e) DNA, and (f) subunit vaccines [36].

Chitosan is a polycationic polymer comprising of D-glucosamine and N-acetyl-D-glucosamine linked by $\beta(1,4)$-glycosides' bonds. It is produced by deacetylation of chitin, which is extracted from the shells of crabs and shrimp. It is a linear, hydrophilic, positively charged, water soluble biopolymer, can form thin films, hydrogels, porous scaffolds, fibers, and micro and nanoparticles in mild acidity conditions. As a polycationic polymer, it has a high affinity to associate macromolecules such as insulin, pDNA, siRNA, heparin, among others, with antigenic molecules, protecting them in turn from hydrolytic and enzymatic degradation [37].

Polyvinylpyrrolidone (N-vinyl-2-pyrrolidone, PVP) has chemical, physical and physiological properties which have been exploited in various industries, including but not limited to medical, pharmaceutical, cosmetic, food, and textile, due to its biological compatibility, low toxicity, tackiness, resistance to thermal degradation in solutions as well as inert behavior in salt and acidic solutions [38, 39]. It is a water soluble homopolymer with a wide range of molecular weights (2.5 to 1.200 kDa), molecules between 12 and 1350 monomers, and end-to-end distances ranging from 2.3 to 93 nanometers. It is physically and chemically stable; it tolerates heating and air atmospheres for up to 16 hours at 100 °C, as well as the change of appearance for 2 months at 24 ° C and 15% HCl. When heated with strong bases such as lithium carbonate, trisodium phosphate or sodium metasilicate, it generates a precipitate due to the ring opening and subsequent crosslinking of chains. Yen-Jen *et al.* studied its effect as a drug deliverer and intracellular acceptor [40].

4.1.1. Molecular docking

In this work, the molecular docking calculations were performed using the AutoDock program. In particular, it uses a Lamarckian genetic algorithm (LGA) and a force field function based approximately on the AMBER force field, which consists of five terms: 1) the 12-6 dispersion term of Lennard-Jones, 2) a 12-10 directional hydrogen bonding term, 3) an electrostatic Coulomb potential, 4) an entropic term, and 5) one term of desolvation pairs. The scaling factor of these terms is empirically calibrated using a set of 30 structurally-known protein-ligand complexes, which affinities have been experimentally determined. The AutoDock program has become widely used due to its good precision and high versatility; moreover, the latest version of AutoDock (version 4.0) added flexible functions to the side chains in the receptor.

4.1.2. Model preparation

In this study we used the E protein of dengue virus. It consists of a dimer with 394 amino acids (aa) per monomer and, as mentioned before, it is the main component of the dengue virus envelope. E protein enters the cell by fusion with the membrane due to a previous conformational change produced by a low pH, generating thereby a change of form from dimer to trimer, in which the fusion peptide between the II and III domains is exposed. When the pH is lower than 6.3, dimers dissociate from dimer phase, making the I and II domains rotate outwards and exposing the fusion loop, which interacts with the endosomal membrane of the cell. Domain III then rotates backwards to pull the I and II domains, which were already bound to the cell membrane by the fusion peptide, thus attaching the cell membrane with the membrane of the virus in order to release the RNA [27, 29, 41-45]. It is important to mention that Bressanelli showed that the virus domains remain at neutral pH but their relative orientation is altered [27]. For best results during the molecular docking process, we optimize the original model of the dengue virus protein E (PDB code 1OKE) with a number of refinements and validations cycles with Phenix and Molprobity programs respectively. Figure 3 shows the corrected model of the dimer and trimer.

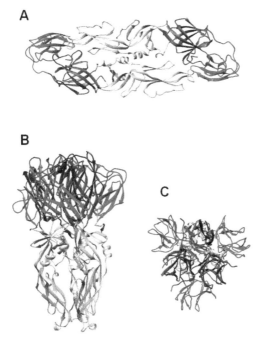

Figure 3. Dengue virus protein E. (A) Dimeric protein after geometric and refining corrections with Phenix program. (B) Trimeric protein that represents a postfusion state (C) Top view of the trimeric form. Domain I (aa$_{1-52,132-193,280-296}$) is in red, domain II (aa$_{52-132, 193-280}$) in yellow and domain III (aa$_{296-394}$) in blue color.

4.1.3. Ligands preparation

To prepare the ligands, we utilized the linear PVP and CS monomers, and CS/TPP/CS chains (Figure 4). The coordinates of those ligands were obtained using the SMILES program [46].

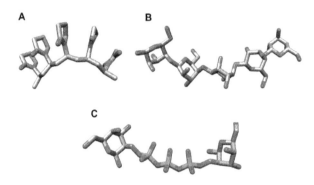

Figure 4. Ligands used in molecular docking. (A) PVP monomer, (B) CS monomer, and (C) CS/TPP/CS chain.

4.1.4. Molecular docking

Molecular docking was performed by means of the AutoDock program that combines rapid grid-based energy evaluation and efficient search of torsional freedom. This program uses a semi-empirical free energy force field to evaluate the conformations during the docking simulations. The force field is quantified using a large number of protein-inhibitor complexes, for which the inhibition constants (K_i), are known. The force field evaluates the union in two steps, first when the ligand and the protein are separated. Then, the intramolecular energies are estimated for the transition from the unbound state to the protein-ligand bound state. In the second step, intermolecular energies are evaluated by combining the ligand with the protein conformations bound to themselves. The force field includes six pairs-wise of evaluations (Vi) and an estimated loss of conformational entropy after binding (ΔS_{conf}):

$$\Delta G = \left(V_{bound}^{L\text{-}L} - V_{unbound}^{L\text{-}L}\right) + \left(V_{bound}^{P\text{-}P} - V_{unbound}^{P\text{-}P}\right) + \left(V_{bound}^{P\text{-}L} - V_{unbound}^{P\text{-}L} + \Delta S_{conf}\right) \tag{1}$$

where L refers to the ligand and P to the "protein" in a ligand-protein docking. Each of the pair-wise energetic terms includes evaluations for dispersion/repulsion, hydrogen bonding, electrostatics, and desolvation [47].

The calculations can be summarized in the following four steps: (1) preparation of files using AutoDockTools coordinates, (2) pre-calculation of atomic affinities by using AutoGrid, (3) docking of ligands by using AutoDock, and (4) analysis of the results applying AutoDockTools.

4.2. Results

4.2.1. *Amino acids of interest in the dengue virus infection mechanism*

In the loop conformation, several amino acids are involved in trimerization of unit E of DENV. These amino acids are of particular interest since they are allocated in between I and II domains, the fusion loop of the host cell located between domains II and III, and aa268-270 (kl loop). Also are important, the loop of fusion to the host cell located between domains II and III, which subsequently is exposed in the trimer with the aa98-111 fusion peptide, and the C-terminal of domain III, which holds the protein to the virus membrane. Other important amino acids were mentioned by Mazumder [48], who made a structural analysis of the dengue virus E protein of the 4 serotypes in order to find the conserved and exposed sites as well as the epitopes in the T-cell. In our study, we additionally considered the sites of interest described by Yorgo Modis [42, 43] (Table 1).

In addition to the ten conserved regions presented in Table 1, we predicted around 740 E proteins of the 4 serotypes, some of which are included in the same Table 1. Their sequences were quantified using Shannon's entropy [48] with a variation from 0.3 to 1.1 bits.

Amino acid	Dimer configuration	Trimer configuration	Function
L191,T268-I270	This region is known as kl loop, without the presence of the ligand (β-Octylglucoside, BOG), it forms a salt bridge and a hydrogen bond with beta strand I and j of the counterpart dimer.	The ligand is present; however, the kl loop does not adopt the open conformation present in the dimer-ligand pair.	A hinge allowing movement of domains I and II, as well as conformational changes when varying the pH of the endosome.
V382-G385	C terminal that attaches domain III to premembrane (prM) virus.	It combines to create a trimeric contact with the other two domains.	It holds and folds the domains I and II, acting as a zipper. It is the most variable region among the 4 serotypes.
D98-G111	The loop is protected between the domain II and III; it contains a fusion peptide that is formed by hydrophobic residues.	It is exposed only in the trimeric conformation during the conformational changes, and it maintains its structure.	It is the region of highest interest since it is the binding receptor for the host cell. In the trimeric form it fuses and binds the cell and virus's membranes. It is the most conserved region among the 4 serotypes.
N37, P207	Exposed.	Exposed.	Conserved epitope in the 4 serotypes.

Amino acid	Dimer configuration	Trimer configuration	Function
Q211 D215	Exposed.	Not exposed.	Conserved epitope in the 4 serotypes.
H244, K246	Not exposed.	Exposed.	Conserved epitope in the 4 serotypes.
C30,121,105,116,285 and 333	30 hidden, 105 semi-exposed, 116 hidden, 121 semi-exposed, 285 semi-exposed, 333 semi-exposed.	All semi-hidden.	Disulfide bonds in the 4 serotypes. It provides structure to the protein.
R9 E368	Salt bridge, conserved region.	Salt bridge, conserved region.	Structural stability, interactions between the domain I and III.
H144,317	Hydrogen bridges, conserved region.	Hydrogen bridges, conserved region.	Structural stability of the main chain with the opposite domain.
N67 N153	Interacts with N acetyl-D-glucosamine (NAG)	Only N67 interacts with NAG.	Glycosylation sites.

Table 1. Sites of interest identified in the structure of the dengue virus E protein according to Yorgo Modis [42, 43, 48].

The analyzed proteins can be identified as: N8-G14, V24-D42, R73-E79, V97-S102, D192-M196, V208-W220, V252-H261, G281-C285, E314-T319, E370-G374, and K394-G399; whereas the hidden amino acids, which change to exposed amino acids in the trimer, can be listed as follows: M1, H244, K246, G254, G330 and K344; and the exposed residues that remain hidden include the following: S16, Q52, Q167, S169, P243, D290, Q293, S331 and E343. It is worth mentioning that we have identified at least 14 conserved negative sites in at least 3 of the 4 serotypes (C3, C60, R73, T189, F213, A267, F306, T319, S376, F392, K394, S424, G445, and V485). The importance of this discovery relies on the fact that it has demonstrated that the epitopes with negative sites work better as vaccines than those with positive sites as they are less likely to change due to their functional restrictions.

4.2.2. Dengue virus E protein — PVP docking

The docking of PVP molecules with the E protein of dengue virus has demonstrated that the interface of the I and II domain was the most energetically favorable site for the binding (Figure 5). The interaction between protein and ligand takes place by establishing 8 hydrogen bonds with the Asn124, Lys202, and Asp203 amino acids (Table 2). This region is extremely important for the pivotal role it plays in the conformational changes triggered by low pH, which in turn is closely related to the infectivity of the virus. In particular, the PVP molecule, which interacts with aa124,202,203 in the E protein-BOG ligand complex, could act as a blocker of the kl aa268-270 pitchfork activity, which is responsible for the conformational changes in the E protein at low pH. In other words, it could inhibit their function to work as a hinge for conformational changes due to its proximity to amino acids through steric hindrance, preventing thereby the hinge action between the I and II domain, which in turn could stiffen the area. Alternatively, if BOG ligand is absent, the molecule could be internalized into the hydrophobic pocket and replace it, but the subsequent molecular prediction simulations would be required to determine how it could act in the presence of low pH, in order to find out whether the conformational changes would appear or be inhibited. The PVP is well-known to be highly stable at acid pH and high temperatures, so its structural integrity is assured to remain intact; the loop or kl pitchfork amino acids mutate, resulting in an increase of the pH threshold, at which conformational changes occur. It is achieved by replacing long hydrophobic side chains by the short ones. As the result, the site can be consistently represented as a potential trigger in the virus replication cycle and a good candidate to inhibit its function (Figure 5).

PVP molecule	Amino acid	Distance (Å)
O1	OD1-Asp 203	2.75
O2	OD1-Asp 203	2.44
O3	N-Asp 203	2.64
O3	N-Lys 202	2.58
O4	N-Lys 202	3.37
O4	O-Asn 124	2.92
O5	O-Asn 124	3.32
O5	N-Asn 124	2.87

Table 2. Polar interactions of the PVP molecule with dengue virus E protein.

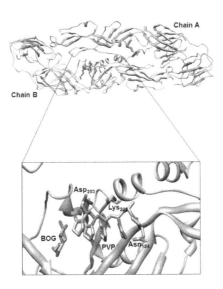

Figure 5. Dengue virus E protein with PVP ligand. The lower part shows a close up of the docking area.

4.2.3. Dengue virus E protein — CS docking

The docking of CS molecules in the E protein of dengue virus resulted in the interaction with the interface of domain I and II of the protein (see Figure 6). The CS ligand binds to seven amino acids of E protein by ten hydrogen bonds (see Table 3). The elongated CS molecule settles into a channel formed in the II domain surface of the protein. Additionally, it interacts with amino acids near the kl hinge or loop of I and II domain interface. There is a remarkable familiarity between the BOG and NAG complexes. Amino acids-CS molecule interactions, which are shown in Table 10 (aa65,68,202,249,251,272,273), suggest that the mechanism of action of this molecule is similar to PVP ligand. Additionally, it is very close to the conserved region V252-H261 that forms a channel in the 4 serotypes. This finding is of the highest importance since it could very well serve as a ligand for the 4 serotypes, and it could be even more useful in the development of a chimera vaccine with the four domains III of E protein, which would be similar to the chimeric vaccine developed in India at the International Centre for Genetic Engineering and Bionanotechnology.

CS molecule atom	Amino acid atom	Distance (Å)
O1 ring 1	O- Lys 202- B	3.47
O1 ring 1	O-Met 272-B	2.97
N ring 2	OG-Ser 273-B	3.24
O1 ring 2	N-Val 251-A	3.14
O1 ring 2	O-Val 251-A	1.97
O3 ring 2	O-Leu 65-A	3.28
O5 ring 2 and O2 ring 3	OD2-Asp 249-A	3.5
O1 ring 3	OG1-Thr 68-A	2.77
N ring 3	OD2-Asp 249-A	2.51
N ring 3	OD1-Asp 249-A	3.33

Table 3. CS molecule interactions with dengue virus E protein.

4.2.4. Dengue virus E protein — CS/TPP/CS docking

In this case, we used the CS and TPP monomers taking into account that the CS units form bindings by means of 1-4 beta bonds. Similarly to the E protein–PVP docking, the molecular docking between the CS/TPP/CS ligand and the E protein was carried out between the domains I and II, although we observed more interactions in the case of PVP monomer. Table 4 and Figure 7 illustrate seven interactions between the amino acids and the BOG. The CS-TPP-CS complex interacts with aa49, 124, 126, 200, 202, 203, 271 amino acids, and the docking results suggest that these three molecules are attracted the most to the area formed by the hydrophobic pocket, indicating that the latter molecule has a direct interaction with the BOG ligand oxygen.

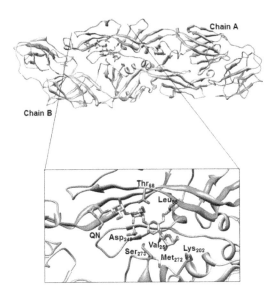

Figure 6. Docking of dengue virus E protein with CS ligand. The interaction takes place at an interface between the two monomers.

CS/TPP molecule atom	Amino acid atom	Distance (Å)
N ring 1	O-Ser 274	3.43
O3 ring 1	OE2-Glu 49	2.90
O5 ring 1	NE2-Gln 271	3.08
O5 ring 1	OE1-Gln 271	2.66
O5 ring 1	O3-BOG	2.74
O1 TPP1	O-Ser274	3.03
O3 TPP1	O4-BOG	3.14
O1 TPP2	OD1-Asp 203	2.83
O3 TPP2	O4-BOG	3.30
O1 TPP3	O-Lys 202	2.19
O1 TPP3	OD1-Asp 203	3.18
O3 TPP3	NE2-Gln 200	3.43
O4 TPP	O-Lys 202	2.51
O1 ring 2	N-Lys 202	2.89
O3 ring 2	ND2-Asn 124	2.89
O3 ring 2	OD1-Asn 124	2.38
O4 ring 2	OE2-Glu 126	2.49

Table 4. Docking of CS/TPP/CS molecule with dengue virus E protein.

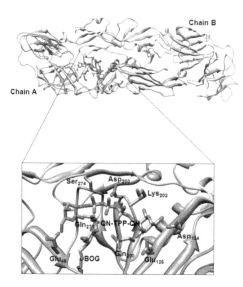

Figure 7. CS/TPP/CS molecule docking with the dengue virus E protein.

5. Conclusions

We have reviewed the key concepts and current experimental procedures, including the recent advances in protein flexibility, ligand sampling, and scoring function. In addition, challenges and possible future directions were addressed in this chapter. As an example of protein ligand study we analyzed the interaction between the dengue virus E protein and Polyvinylpyrroli-done and Chitosan biopolymers and we confirmed that PVP, CS, and CS/TPP/CS biopolymers can fulfill the function of adjuvant carriers in the potential development of a chimeric dengue vaccine against the 4 serotypes of dengue virus. Furthermore, the ring-shaped molecules have shown affinity to or preference for a place of vital importance in the virus's cycle of infection and replication, which placed us on the path to develop an inhibitor of the aforementioned conformational changes (see Figures 5-7). Their binding to the E protein is possible due to the great affinity they present to simulated molecules. However, further analysis of molecular simulation is required to determine the behavior of the protein without the presence of BOG ligand or in different environmental conditions in the presence of low pH.

Acknowledgements

This work has been supported by Fomix-Veracruz (2009-128001) and CONACyT-Mexico (CB2008-105491, CMB).

Author details

Alejandra Hernández-Santoyo[1], Aldo Yair Tenorio-Barajas[2], Victor Altuzar[2], Héctor Vivanco-Cid[3] and Claudia Mendoza-Barrera[2*]

*Address all correspondence to: omendoza@uv.mx

1 Instituto de Química, Universidad Nacional Autónoma de México, Mexico, D.F., Mexico

2 Laboratorio de Nanobiotecnología, Centro de Investigación en Micro y Nanotecnología, Universidad Veracruzana, Boca del Rio, Veracruz, Mexico

3 Instituto de Investigaciones Médico-Biológicas, Universidad Veracruzana, Boca del Río, Veracruz, Mexico

References

[1] Huang, S. Y, & Zou, X. Advances and challenges in protein-ligand docking. International Journal of Molecular Sciences (2010). , 11(8), 3016-3034.

[2] Halperin, I, Ma, B, Wolfson, H, & Nussinov, R. Principles of docking: An overview of search algorithms and a guide to scoring functions. Proteins (2002). , 47(4), 409-443.

[3] Sousa, S. F, Fernandes, P. A, & Ramos, M. J. Protein-ligand docking: current status and future challenges. Proteins (2006). , 65(1), 15-26.

[4] Wang, R, Fang, X, Lu, Y, Yang, C. Y, & Wang, S. The PDBbind Database: Methodologies and updates. Journal of Medicinal Chemistry (2005). , 48(12), 4111-4119.

[5] Puvanendrampillai, D, & Mitchell, J. B. L. D Protein Ligand Database (PLD): additional understanding of the nature and specificity of protein-ligand complexes. Bioinformatics (2003). , 19(14), 1856-1857.

[6] Block, P, Sotriffer, CA, Dramburg, I, Klebe, G, & Affin, . : a freely accessible database of affinities for protein-ligand complexes from the PDB. Nucleic Acids Research 2006; 34 D522-526.

[7] Liu, T, Lin, Y, Wen, X, Jorissen, RN, Gilson, MK, & Binding, . : a web-accessible database of experimentally determined protein-ligand binding affinities. Nucleic Acids Research 2007; 35 D198-201.

[8] Dias, R. de Azevedo WF Jr. ((2008). Molecular docking algorithms. Curr Drug Targets. Dec; , 9(12), 1040-7.

[9]   Kuntz, I. D, Blaney, J. M, Oatley, S. J, Langridge, R, & Ferrin, T. E. A geometric approach to macromolecule-ligand interactions. Journal of Molecular Biology (1982). , 161(2), 269-88.

[10] Jain, A. N. Scoring functions for protein-ligand docking. Current Protein Peptide Science (2006). , 7(5), 407-20.

[11] Gilson, M. K, & Zhou, H. X. Calculation of protein-ligand binding affinities. Annual Review of Biophysics and Biomolecular Structure (2007). , 36-21.

[12] Huang, S. Y, Grinter, S. Z, & Zou, X. Scoring functions and their evaluation methods for protein-ligand docking: recent advances and future directions. Physical Chemistry Chemical Physics (2010). , 12(40), 12899-908.

[13] Taylor, R. D, Jewsbury, P. J, & Essex, J. W. A review of protein-small molecule docking methods. Journal of Computer-Aided Molecular Design. (2002). , 16(3), 151-66.

[14] Chen, V. B, & Arendall, W. B. rd, Headd JJ, Keedy DA, Immormino RM, Kapral GJ, Murray LW, Richardson JS, Richardson DC. MolProbity: all-atom structure validation for macromolecular crystallography. Acta Crystallographica Section D: Biological Crystallography. (2010). Pt 1):12-21.

[15] Morris, G. M, Huey, R, Lindstrom, W, Sanner, M. F, Belew, R. K, Goodsell, D. S, & Olson, A. J. AutoDock4 and AutoDockTools4: Automated docking with selective receptor flexibility. Journal of Computational Chemistry (2009). , 16-2785.

[16] Rarey, M, Kramer, B, Lengauer, T, & Klebe, G. A fast flexible docking method using an incremental construction algorithm. Journal of Molecular Biology (1996). , 261(3), 470-89.

[17] Jones, G, Willett, P, Glen, R. C, Leach, A. R, & Taylor, R. Development and validation of a genetic algorithm for flexible docking. Journal of Molecular Biology (1997). , 267(3), 727-748.

[18] Abagyan, R, Totrov, M, & Kuznetsov, D. ICM-A new method for protein modeling and design: Applications to docking and structure prediction from the distorted native conformation. Journal of Computational Chemistry (1994). , 15(5), 488-506.

[19] Mizutani, M. Y, Tomioka, N, & Itai, A. Rational automatic search method for stable docking models of protein and ligand. J. Mol. Biol. (1994). , 243, 310-326.

[20] Taylor, J. S, & Burnett, R. M. DARWIN: A program for docking flexible molecules. Proteins (2000). , 41, 173-191.

[21] Clark, K. P. Flexible ligand docking without parameter adjust-ment across four ligand-receptor complexes. J. Comput. Chem. (1995). , 16, 1210-1226.

[22] Hart, T. N, & Read, R. J. A multiple-start Monte Carlo docking method. Proteins (1992). , 13, 206-222.

[23] World Health OrganizationWHO: Media Centre, Fact sheets: Dengue and severe dengue. http://www.who.int/accessed 15 August (2012).

[24] Huang, J. H, Wey, J. J, Sun, Y. C, Chin, C, Chien, L. J, & Wu, Y. C. Antibody responses to an immunodominant nonstructural 1 synthetic peptide in patients with dengue fever and dengue hemorrhagic fever. Journal of Medical Virology (1999). , 57(1), 1-8.

[25] Muñoz, M. L, Cisneros, A, Cruz, J, Das, P, Tovar, R, & Ortega, A. Putative dengue virus receptors from mosquito cells. FEMS Microbiology Letter (1998). , 168(2), 251-258.

[26] Cao-lormeau, V. M. Dengue viruses binding proteins from Aedes aegypti and Aedes polynesiensis salivary glands. Virology Journal (2009).

[27] Bressanelli, S, et al. Structure of a flavivirus envelope glycoprotein in its low-pH-induced membrane fusion conformation. The EMBO Journal (2004). , 23(4), 728-738.

[28] Mondotte, J. A, Lozach, P. Y, Amara, A, & Gamarnik, A. V. Essential Role of Dengue Virus Envelope Protein N Glycosylation at Asparagine-67 during Viral Propagation. Journal Virology (2007). , 81(13), 7136-7148.

[29] Mukhopadhyay, S, Kuhn, R. J, & Rossman, M. G. A structural perspective of the flavivirus life cycle. Nature Reviews Microbiology (2005). , 13-22.

[30] Stiasny, K, Allison, S. L, Schalich, J, & Heinz, F. X. Membrane Interactions of the Tick-Borne Encephalitis Virus Fusion Protein E at Low pH. Journal of Virology (2002). , 76(8), 3784-3790.

[31] Mandl, C. W, Guirakhoo, F, Holzmann, H, Heinz, F. X, & Kunz, C. Antigenic structure of the flavivirus envelope protein E at the molecular level, using tick-borne encephalitis virus as a model. Journal of Virology (1989). , 63(2), 564-571.

[32] Acosta-bas, C, & Gómez-cordero, I. Biología y métodos diagnósticos del dengue. Revista Biomedica (2005). , 16-113.

[33] Kelly, E. P, Greene, J. J, King, A. D, & Innis, B. L. Purified dengue 2 virus envelope glycoprotein aggregates produced by baculovirus are immunogenic in mice. Vaccine (2000). , 18(23), 2549-2559.

[34] Putnak, R, et al. Immunogenic and protective response in mice immunized with a purified, inactivated, Dengue-2 virus vaccine prototype made in fetal rhesus lung cells. The American Journal of Tropical Medicine Hygiene (1996). , 55(5), 504-510.

[35] Staropoli, I, Grenckiel, M. P, Mégret, F, & Deubel, V. Affinity-purified dengue-2 virus envelope glycoprotein induces neutralizing antibodies and protective immunity in mice. Vaccine (1997).

[36] Heinz, F. X, & Stiasny, K. Flaviviruses and flavivirus vaccines. Vaccine, (2012). , 30(29), 4301-4306.

[37] Sundar, S, Kundu, J, & Kundu, S. C. Biopolimeric nanoparticles. Science and Tech-
      nology of Advanced Materials (2010).

[38] Robinson, B. V, Sullivan, F. M, Borzelleca, J. F, & Schwart, S. L. PVP : a Critical Re-
      view of the Kinetics and Toxicology of Polyvinylprrolidone (Povidone). Chelsea, MI:
      Lewis Publishers; (1990).

[39] Buhler, V. Polyvinylpyrrolidone Excipients for Pharmaceuticals: Povidone, Crospovi-
      done, and Copovidone. Berlin, New York: Springer-Verlag; (2005).

[40] Wang, Y. J, Chien, Y. C, Wu, C. H, & Liu, D. M. Magnolol-loaded core-shell hydrogel
      nanoparticles: drug release, intracellular uptake, and controlled cytotoxicity for the
      inhibition of migration of vascular smooth muscle cells. Molecular Pharmaceutics
      (2011). , 8(6), 2339-2349.

[41] Harrison, S. C. The pH sensor for flavivirus membrane fusion. The Journal of Cell Bi-
      ology (2008). , 183(2), 177-179.

[42] Modis, Y, Ogata, S, Clements, D, & Harrison, S. C. Structure of the dengue virus en-
      velope protein after membrane fusion. Nature (2004). , 427-313.

[43] Modis, Y, Ogata, S, Clements, D, & Harrison, S. C. A ligand-binding pocket in the
      dengue virus envelope glycoprotein. Proceeding of the National Academic Science of
      the United State of America (2003). , 100(12), 6986-6991.

[44] Yu, I. M, Holdaway, H. A, Chipman, P. R, Kuhn, R. J, Rossmann, M. G, & Chen, J.
      Association of the pr Peptides with Dengue Virus at Acidic pH Blocks Membrane Fu-
      sion. Journal of Virology (2009). , 83(23), 12101-12107.

[45] Sánchez- San Martin CLiu CY, Kielian M. Dealing with low pH: entry and exit of al-
      phaviruses and flaviviruses. Trends in Microbiology (2009). , 17(11), 514-521.

[46] Weininger, D. SMILES, a Chemical Language and Information System. 1. Introduc-
      tion to Methodology and Encoding Rules. Journal of Chemical Information and
      Computational Science (1998). , 28(1), 31-36.

[47] Garrett, M, Morris, D. S. G, & Michael, E. Pique, William "Lindy" Lindstrom, Ruth
      Huey, Stefano and W.E.H. Forli, Scott Halliday, Rik Belew and Arthur J. Olson, User
      Guide AutoDock Version 4.2, Automated Docking of Flexible Ligands to Flexible Re-
      ceptors. (2010). , 49.

[48] Mazumder, R, Hu, Z. Z, Vinayaka, C. R, & Sagripanti, J. L. Frost SDW, Kosakovsky
      P, Wu CH. Computational analysis and identification of amino acid sites in dengue E
      proteins relevant to development of diagnostics and vaccines. Virus Genes (2007). ,
      35(2), 175-186.

Chapter 3

Bioprocess Engineering of *Pichia pastoris*, an Exciting Host Eukaryotic Cell Expression System

Francisco Valero

Additional information is available at the end of the chapter

1. Introduction

Yeasts are the favorite alternative hosts for the expression of heterologous proteins for research, industrial or medical use [1]. As unicellulars microorganism have the advantages of bacteria as ease of manipulation and growth rate. But comparing with bacterial system, they are capable of many of the post-translational modifications performed by higher eukaryotic cells, such as proteolytic processing, folding, disulfide bond formation and glycosylation [2].

Historically *Saccharomyces cerevisiae* has been the most used yeast host due to the large amount of knowledge on genetics, molecular biology and physiology accumulated for this microorganism [3-5]. However, it was rapidly found to have certain limitations: low product yields, poor plasmid stability, hyperglycosylation and low secretion capacities. These limitations are now relieved by a battery of alternative yeast as cell factories to produce recombinant proteins.

Some of these alternative yeast cell factories are fission yeast as *Schizosaccharomyces pombe* [6], *Kluyveromyces lactis* [7], methylotrophic species as *Pichia pastoris* [8], *Candida boidinii* [9], *Pichia methanolica* [10], *Hansenula polymorpha* [11], and the dimorphic species *Yarrowia lipolytica* [12], and *Arxula adeninivorans* [13]. It is very usual that the performance of these alternative hosts frequently surpass those of *S. cerevisiae* in terms of product yield, reduced hyperglycosylation and secretion efficiency, especially for high molecular weight proteins [14].

Several reviews compare advantages and limitations of expression systems for foreign genes [15-20]. Between them *Pichia pastoris* has emerged in the last decade as the favorite yeast cell factory for the production of heterologous proteins. A search in ISI Web of knowledge (web of science) with the keywords microorganism+ heterologous protein *P. pastoris* is the preferred host (667 entrances) followed by *Candida* and *Schizosaccharomyces*

(161 and 124 entrances respectively). Specifically for heterologous lipase production *P. pastoris* is the most used host [21].

Why *P. pastoris* emerged as an excellent host system to produce recombinant products?. The story started one decade after oil crisis in the 70's when Phillips Petroleum and the Salk Institute Biotechnology/Industrial Associates Inc. (SIBIA, La Jolla, Ca, USA) used *Pichia* as a host system for heterologous protein expression [22-24]. Nowadays, more than 500 proteins have been expressed using this system [25] and it also has been selected by several protein production platforms for structural genomics programs [26]. *P. pastoris* combines the ability of growing on minimal medium at very high cell densities (higher than 100 g DCW/L), secreting the heterologous protein simplifying their recovery. Also, it performs many of the higher eukaryotic post-translational modifications such as protein folding, proteolytic processing, disulfide bond and glycosylation [24]. However, it has been shown that both, N- and O-linked oligosaccharide structures, are quite different from mammalian cells, for example, they are of a heterogeneous high-mannose type. The consequence is that high mannose type N-glycans attached to recombinant glycoproteins can be cleared rapidly from the human bloodstream, and they can cause immunogenic reactions in humans [27]. Nevertheless GlycoFi's glyco-engineering technology allows the generation of yeast stains capable of replicating the most essential steps of the N-glycosylation pathway found in mammals [28].

But, probably the most important characteristic of *P. pastoris* is the existence of a strong and tightly regulated promoter from alcohol oxidase 1 gene, *PAOX1*. Thus, methanol was used as carbon source and inducer of heterologous protein production in this system [29].

Daly and Hearn [30] reviewed various aspects of the *P. pastoris* expression system and also consider the factors that need to be taken into account to achieve successful recombinant protein expression, particular when more complex systems are contemplated, such as those used in tandem gene or multiple gene copy experiments. Between them, several genetic and physiological factors such as the codon usage of the expression gene, the gene copy number, efficient transcription by using strong promoters, translation signals, translocation determined by the secretion signal peptide, processing and folding in the endoplasmatic reticulum and Golgy and, finally, secretion out of the cell, as well as protein turnovers by proteolysis, but also of the optimization of fermentation strategy [31].

The objective of this chapter is to review the classic and alternative operational strategies to maximize yield and/or productivity from an industrial point of view and also how to obtain a repetitive product from batch to batch applying process analytical technology (BioPAT)

2. Host strains and *PAOX1* promoter

Host strains and vectors are available as commercial kits from Invitrogen Corporation (Carlsbad, CA) [32]. *PAOX1* is the preferred promoter. Previous to design operational strategies is necessary to know the machinery to inducer this promoter and how *Pichia* metabolizes methanol.

PAOX1 is strongly repressed in presence of carbon sources as glucose, glycerol, ethanol and most of other carbon sources, being strongly induced by the presence of methanol [33]. Alcohol oxidase is the first enzyme of methanol assimilation pathway, which catalyzes its oxidation to formaldehyde [34]. The genome of *Pichia* contains two genes of this functional enzyme AOX1 and AOX2. Around the 85% of alcohol oxidase activity is regulated by *AOX1* gene, whereas *AOX2* gene regulates the other 15% [35]. AOX concentration can reach 30% of the total cell protein when is growing on methanol, which compensates for the low affinity of the enzyme for methanol [22].

There are three types of *P. pastoris* host strains available which vary with regards to their ability to utilize methanol. The wild-type or methanol utilization plus phenotype (Mut$^+$), and the strains resulting from deletions in the *AOX1* gene, methanol utilization slow (Mut^s), or both *AOX* genes, methanol utilization minus (Mut$^-$) [36].

Although AOX1 is the promoter most commonly used, it presents a serie of limitations. Oxygen supply becomes a major concern in *P. pastoris* in methanol non-limited fed-batch cultures when high cell densities are desired for the production process using Mut$^+$ phenotype, since the bioreactor oxygen transfer capacity unable to sustain the oxygen metabolic demand [24]. Another important disadvantage of *PAOX1*, especially in Mut$^+$ phenotype in large scale productions, is the necessity to storage huge amount of methanol which constitutes a potential industrial risk. On the other hand, methanol presents a high heat of combustion (-727 kJ C-mol^{-1}) [37]. Thus, considerable heat is generated during the bioprocess growing on this carbon source. It requires rapid and efficient cooling systems, particularly at large scale where heat losses through the bioreactor walls may be limiting due to the small surface area to volume ratio. Failure to remove this heat may result in reactor temperature increase affecting the productivity and quality of the recombinant protein [38]. Furthermore since methanol is mainly derived from petrochemical sources, may require purifications steps for the production of certain foods and additives products [39].

3. *Pichia* Process Analytical Technology (PAT)

It is necessary to develop bioprocess optimization and control tools in order to implement a Process Analytical Technology (PAT), BIOPAT when it is applied to bioprocesses [40]. This initiative has been promoted by regulatory agencies such as FDA and EMEA [41]. PAT is a multidisciplinary platform for designing, analyzing and controlling manufacturing through timely measurements of critical quality and performance attributes of raw and in-process materials and processes with the goal of ensuring final product quality [42].The final goal is guarantee consistent product quality at the end of the process, ease the regulatory review bioprocess and increase flexibility with respect to post-approval manufacturing changes [43] [Figure 1].

Applied to *Pichia* cell factory, on-line monitoring of biomass, methanol and product are the dream of all researchers involved in the production of heterologous protein in this host.

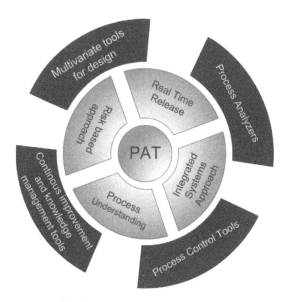

PAT - Process Analytical Technology

Figure 1. Scheme of a process analytical technology (PAT).

Different approaches have been applied for the on-line determination of biomass in *Pichia's* fermentation. Multi-wavelength fluorescence coupling with PARAFAC-PLS chemometric methodology resulting in important qualitative and quantitative bioprocess information [Figure 2; Figure 3]. Biomass and substrate (glycerol or methanol) were determined successfully. The recombinant lipase, the heterologous product, could also be on-line determined in the exponential phase. However in the stationary Phase, where proteolytic problems appears, the estimation of the product could not be estimated accurately [44-46].

Multi-wavelength fluorescence is not standard equipment used in bioprocesses. Thus, when direct biomass quantification methods are not available, biomass can be determined from indirect on-line measurements using software sensors. The estimation of biomass, substrate and specific growth rate by two non-linear observers, nonlinear observed-based estimator (NLOBE) and second-order dynamic tuning (AO-SODE) and a linear estimator, recursive least squares with variable forgetting factor (RLS-VFF) have been applied to *Pichia* bioprocess using different indirect measurements, carbon dioxide transfer rate (CTR), oxygen uptake rate (OUR) from conventional infrared and paramagnetic gas analysis, and sorbitol. The AO-SODE algorithm using OUR on-line measurement was the most efficient approach demonstrating the robustness of this methodology [47]. A comparison of the performance of the different observers is presented in table 1.

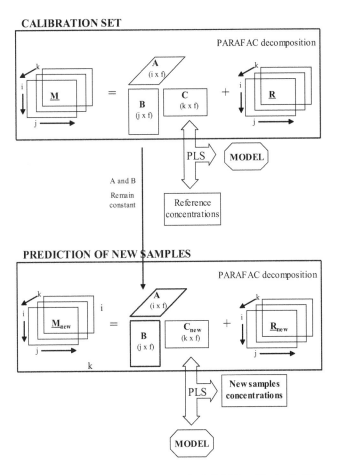

Figure 2. Scheme of the calibration and prediction processes for PARAFAC combined with PLS regression for state variables determination.

Methanol concentration, the inducer substrate, is the most important variable for on-line monitoring because the productivity of the bioprocess is quite related to this parameter. Concentrations between 2-3.5 g/L are referenced as optimal concentrations to maximize protein production [48,49], higher concentrations present inhibition problems and in some cases lower concentration stops recombinant protein production [50].

Although chromatographic methods such as GC and HPLC are common methods for the off-line analysis, their on-line implementation is not usual due to the low sampling frequency [49].

On-line methods are generally based on liquid-gas equilibrium by analyzing the fermenter exhaust gas [51]. Nowadays, commercial equipments based in this principle are available from

a) Creating the PARAFAC model with a NOC batch

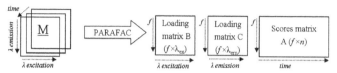

b) Monitoring the process at *n* times

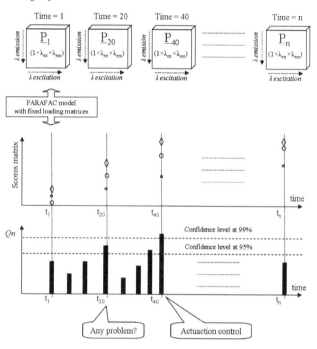

Figure 3. Summary of the application of on-line PARAFAC approach (NOC = Normal Operating Conditions).

Raven Biotec, Figaro Biotech, PTI Instruments [52]. These equipments are quite robust and with minimum maintenance although some precautions should be taking into account to obtain a precise measurement [53].

Other alternatives are sequential injection analysis [54] Fourier transform mid-infrared spectroscopy [49] and flame ionization [55].

Process optimization only can conclude with effective measurement of heterologous protein production. Classical methods as ELISA, SDS-PAGE and Western blots or bioactivity assay are time-consuming, labour-intensive, and not applicable for the determination of the product in real time [51]. Methods including perfusion chromatography, specific biosensors and

Methods	Advantages	Disadvantages
NLOBE	Easy tuning, 1 tuning parameter.	Strong dependence of initial values and kinetic yields.
AO-SODE	Rapid and stable response. Easy tuning, 2 tuning parameters.	Accurate knowledge of reaction scheme and stoichiometric coefficients are necessary.
RLS-VFF	Minimal knowledge of the system.	Sensible to rapid changes of μ.

Table 1. Comparison of three different observers for the estimation of biomass, substrate and specific growth rate.

immunonephelometric assays are limited to proteins secreted in the extracellular culture broth, but not intracellular protein production [56,57]. To circumvent this problem fusing a GFP signal marker to the recombinant protein could be detected by fluorescence [58]. However the co-expression of this protein fusion could provoke a lost in the production of the recombinant product. When the recombinant protein has an associated colorimetric reaction, for instance enzymes, analytical approaches using flow injection analysis (FIA) or sequential injection analysis (SIA) are widely used [59].One of the most fully automated *Pichia* bioprocess has been developed by the group of professor Luttmann [60]. An example of on-line monitoring and control of *Pichia* bioprocess producing *Rhizopus oryzae* lipase is presented in Figure 4. The real time evolution of the main parameters, variables and specific rates of this bioprocess are presented in Figure 5a and 5b.

4. Operational strategies using *PAOX1* Mut⁺ phenotype

Some of the operational strategies using the phenotype Mut⁺ are focused in order to circumvent operational problems previously commented. Invitrogen Co., only provides an operational manual for the fed-batch growth on *Pichia* (Manual Invitrogen) [61] mainly derived from the protocols of Brierley and coworkers [62]. Fed-batch fermentation protocols include three different phases. A glycerol batch phase (GBP), a transient phase (TP) and finally, a methanol induction phase (MIP). Normally GBP and TP are similar for both phenotypes (Mut⁺ and Mutˢ). The objective of the GBP is the fast generation of biomass previous to the induction of methanol. The specific growth rate and yield of *Pichia* growing on glycerol are from 0.18 h⁻¹ and 0.5 g DCW per gram of glycerol [63] to 0.26 h⁻¹ and 0.7 g DCW per gram of glycerol [67]. Brierley and coworkers recommended a maximum glycerol concentration of 6% [62]. Higher concentration inhibits growth [68]. The specific growth rate and yield is higher than growing on methanol (0.12 h⁻¹) and 0.27 g DCW per gram of methanol) [62]. When higher initial biomass concentration is required a second step with an exponential feeding rate of glycerol is implemented. It is important that in GBP dissolved oxygen (DO) reaches values higher than 20-30% to avoid the production of ethanol.

Once the GBP is finished, indicated by a spike in measured DO or a decreased in CO_2 consumption rate (CER), TP is started. The objective of TP is increase biomass level to generate

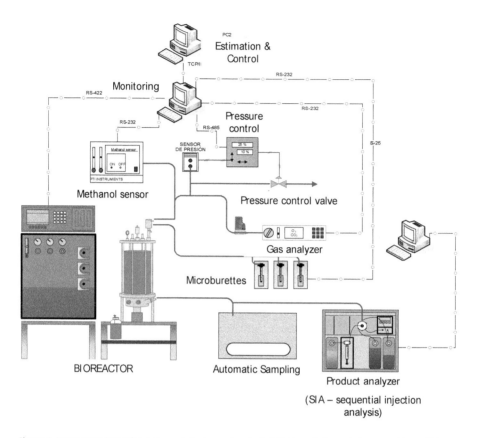

Figure 4. Bioprocess scheme for on-line monitoring and control of *Pichia pastoris* producing recombinant *Rhizopus oryzae* lipase.

high cell density cultures jointly with the derepression of AOX1 promoter due to the absence of an excess of glycerol prior to MIP. Different strategies are collected in a set of reviews [32, 34, 51, 52].

The selected operational strategy used in the MIP is one of the most important factors to maximizing heterologous protein production [67]. These strategies using a Mut$^+$ phenotype have to circumvent the associated problems to the maximum methanol consumption capacity previously pointed out.

At his point, the monitoring and control of the inducer substrate, methanol, are the most important key parameter. High levels of this inducer substrate can generate inhibitory effect on cell growth [67], and low levels of methanol may not be enough to initiate the AOX transcription [8]. The inhibition profile on methanol follows an uncompetitive inhibition growth model, with a reported critic methanol concentration between 3 and 5.5 g/L depending

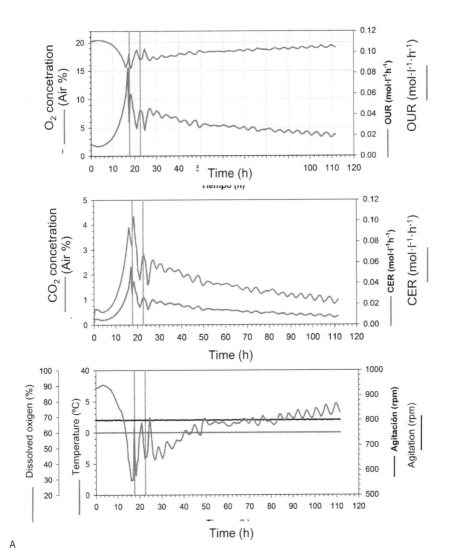

A

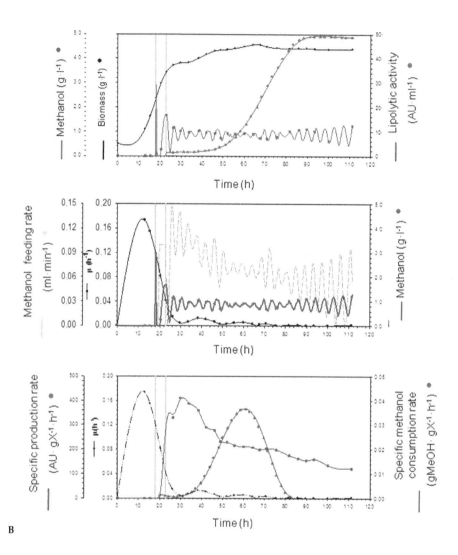

B

Figure 5. A.- Example of the on-line monitoring of *Pichia pastoris* producing recombinant *Rhizopus oryzae* lipase. Real time performance of standard fermentation parameters. B.- Example of the on-line monitoring of *Pichia pastoris* producing recombinant *Rhizopus oryzae* lipase. Real time evolution of biomass, substrate and product and their corresponding specific growth rate, methanol consumption rate and lipase production rate.

on the target protein [34]. Thus, a set-point methanol concentration around 2 g/L seems an optimal value to maximize protein production. Although keeping a constant methanol concentration during the induction phase has positive effects on the production of foreign protein [65], some authors pointed out that the design of an optimal methanol or specific growth rat profile along the MIP maximize the productivity of the process [68].

It is quite difficult to compare the performance of different fed-batch strategies with different heterologous protein. On the other hand, the selection of the fed-batch strategy depends on the facilities to monitor methanol or other key variables as biomass or recombinant product.

Simple strategies, like the addition of a pulse of methanol at different time intervals, must be limited in basic studies to obtain a quantity of recombinant protein for preliminary characterization or structural studies, but is not realistic from an industrial point of view.

Several strategies have been proposed to optimize the methanol feeding rate with the final objective of maximizing protein production and to get a reproducible bioprocess:

5. DO-stat control

Pichia cells utilize methanol through the oxidative pathway only when oxygen is non-limiting [34]. Thus, DO must be controlled above a minimal level around 20% [69]. However, oxygen limitation was successfully used to control the methanol uptake during single-chain antibody fragment production [70,71] and other groups have proposed using oxygen as the growth-limiting nutrient, instead of methanol to circumvent the problem of high oxygen demand and observed 16-55% improvements in product concentrations [72,73]. Recently, an oxygen-limited process has been developed and optimized for the production of monoclonal antibodies in glycoengineered *P. pastoris* strain using oxygen uptake rate as a scale-up parameter from 3L laboratory scale to 1200 L pilot plant scale. Scalability and productivity were improved reducing oxygen consumption and cell growth [74-76]. On the other hand, excessive high DO levels are cytotoxic reducing cell viability [77].

Although different DO-start control has been developed [77-80]. This strategy cannot distinguish the possible accumulation of methanol. In this situation DO signal increases due to the inhibitory effect of methanol on growth, and the response of the DO-controller should be to increase the feeding rate of methanol aggravating the problem. This is particularly problematic in the beginning of the induction phase where *AOX1* is not yet strongly induced and the AOX activity in the cells is growth-rate limiting but constantly increasing as a result of the induction [32].

5. Methanol open-loop control

In this simple strategy, the methanol feeding rate profile (exponential) is obtained from mass balance equations with the objective to maintain a constant specific growth rate (μ) under methanol limiting conditions (no accumulation of methanol should be observed). To implement preprogrammed exponential feeding rate strategy, biomass concentration and volume at the beginning of the MIP have to be known and to assume that a constant biomass/substrate yield is maintained along the induction phase. This strategy has problems in terms of robustness and process stability, because, although open-loop system could be easy to implement they do not respond to perturbations of the bioprocess. To avoid this problem the set point of μ is fixed far from the μ_{max} diminishing the productivity of the process. Nevertheless this simple strategy has been applied successfully in different bioprocesses [81-84]. On the other hand, when the recombinant protein affects the growth of the host reaching μ_{max} lower than the wild strain, like in the production of *Rhizopus oryzae* lipase under methanol limiting conditions, the production is stopped few hours later of the beginning of MIP (personal communication of the author).

6. Methanol closed-loop control

In previous strategies methanol concentration is neither measured on-line not directly controlled [51]. Thus, an accurate monitoring and control of methanol concentration is required. As previously has been commented, different analytical approaches has been implemented in order to on-line monitoring of methanol concentration in *Pichia's* fermentation. Analytical devices based on liquid-gas equilibrium by analyzing exhaust gas from the fermented are the most used. There are as set of methanol sensors available in the market from Raven Biotech, Figaro Electronics, PTI Instruments, and Frings America [52, 85]. The first attempts have been based to maintain the methanol concentration along the induction phase at a constant and optimal concentration to maximize protein production or productivity bioprocess. However, in the last years, some approaches are implementing in order to define an optimal variable methanol set-point function of the different stages of the induction phase. A scheme of both methanol feeding strategies, open and closed loop, is presented in Figure 6.

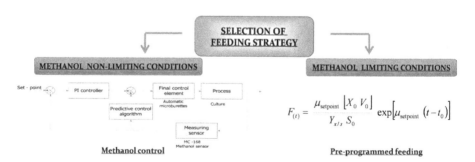

Figure 6. Scheme of methanol feeding strategies: open loop and closed loop control.

Different methanol control concentration algorithms and strategies have been proposed. Although the on-off control is the simplest feed-back control strategy and it has been used by different authors [81, 85-88] *Pichia* fermentation, as bioprocesses in general, is characterized by a complex and highly non-linear process dynamics. For this reason this control strategy is inadequate for precise control of methanol concentration in the bioreactor because it can result

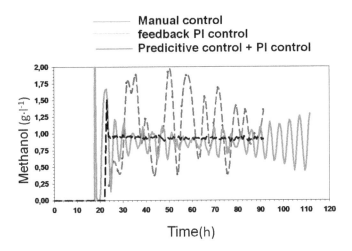

Figure 7. Comparison of the performance of the different methanol control algorithms in *Pichia pastoris* bioprocess producing recombinant lipase.

in a fluctuating methanol concentration around the set-point [34]. In Mut^s phenotype, where the methanol consumption rate is lower than in Mut$^+$ phenotype, this control algorithm has better performance.

A proportional-integral (PI) or proportional-integral derivative (PID) control algorithms are more effective approach. Nevertheless, the optimal settings of the PID controller (gain K_C, the integral time constant τ_I and the derivative time constant τ_D) are hardly ascertained by trial and error tuning or other empirical methods. Some authors have developed a PID control Bode stabilization criterion to achieve the parameters associated to this king of control, obtaining good results on methanol regulation in short time fermentations [77,88]. Because of the dynamics of the system, the optimal control parameters may vary significantly during the fermentation. Moreover, the existence of an important response time for both, the on-line methanol determination and the biological system has promoted the development of other control alternatives [34].

A predictive control algorithm coupled with a PI feedback controller has been implemented successfully in heterologous *Rhizopus oryzae* lipase production. It is based on the methanol uptake on-line calculation from the substrate mass balance in fed-batch cultivations, requiring the first-time derivative of methanol concentration for each time interval. This predictive part is coupled to a feed-back term (PI) to regulate the addition aiming a stabilizing the signal around the set-point [89]. Although this strategy was implemented in Mut^s phenotype, it has been implemented in Mut$^+$ phenotype successfully. A comparison of the performance of the different control algorithms is presented in Figure 7.

Model based on-line parameters estimation and on-line optimization algorithms have been developed to determine optimal inducer feeding rates. Continuous fermentation using methanol was performed via on-line methanol measurement and control using a minimal-variance-controller and a semi-continuous Kalman-Filter [90].

7. Strategies to minimize oxygen demand

The standard fed-batch fermentation without oxygen limitation is namely methanol non-limited fed-batch (MNLFB). Independently of the strategy selected, high cell density cultures with Mut⁺ *P. pastoris* phenotype in laboratory bioreactors presents the problems of oxygen supply, since the bioreactor oxygen transfer capacity is unable to sustain the oxygen metabolic demand [91]. When the biomass reaches values higher than 60 gDCW/L oxygen limitations appears, even using mixtures of air and oxygen or pure oxygen. Different approaches have been published to overcome this disadvantage:

Temperature-limited fed-batch (TLFB). In this strategy the common methanol limitation is replaced by temperature limitation in order to avoid oxygen limitation at high cell density limitation [92]. Temperature controller was programmed to maintain a DO set-point around 25%,. When DO is lower than the set-point the culture temperature was decreased [32]. Using this approach cell death values decrease drastically and also protein proteolysis where reduced, although specific growth rate diminishes and, sometimes, it affect negatively to the productivity of the bioprocess [92]. This strategy has been applied successfully in different heterologous protein production [92-96].

Methanol limited fed-batch strategy (MLFB). The strategy is applied once the DO value under non limited conditions achieves values lower than the set-point (around 25%). At this point methanol feeding rate is controlled in order to assure the DO set-point. At this point methanol concentration starts to diminish from the methanol set-point to limiting conditions, although specific productivity can diminish the production of the heterologous product is not stopped [84, 91, 97-98].

8. Operational Strategies using *PAOX1* Mutˢ Phenotype

Probably Mut⁺ phenotype under *PAOX1* is the most common *P. pastoris* strain used. However, as it has been commented along the chapter, it presents important operational problems related to oxygen and heat demand and methanol security requires. From the biological point of view, Mutˢ phenotype can be used, since they require less oxygen supply and heat elimination. However, the specific growth rate using methanol as sole carbon source is too low compared with Mut⁺, and low levels of biomass are produced [34,50]. Although from the bioprocess engineering point of view the slow operational conditions facilitates the control and reproducibility of the bioprocess, the fermentation time increase and sometimes the productivity of the process decreases drastically.

9. Mixed substrates

All the strategies previously described for Mut$^+$ phenotype can be applied to Mut^s phenotype, but to increase cell density and process productivity, as well as to reduce the induction time, a typical approach is the use of a multicarbon substrate in addition to methanol. It is a simple strategy to increase the energy supply to recombinant cells and the concentration of the carbon sources in the culture broth [81, 86, 88].

One of the most selected substrates is glycerol. Several authors have reported that the use of mixed feeds of glycerol and methanol during the induction phase increase productivity and feeding rates [99]. The advantages to use glycerol as co-substrate is that enthalpy of combustion of glycerol -549,5 kJC-mol^{-1} [100] is lower than the enthalpy of combustion of methanol, -727 kJC-mol^{-1} [37]. Thus, less heat will be released using mixing substrates compared with methanol alone. On the other hand, oxygen consumption is also reduced since less oxygen is necessary for the oxidation of glycerol [38]. Therefore, any method which reduces the heat and oxygen consumption rate without affecting productivity would clearly advantageous.

However, glycerol is reported to repress the expression of alcohol oxidase and subsequently the expression of the target protein [101]. Thus, the rational design of operational strategies for the addition of both substrates in fed-batch fermentation, while avoiding glycerol repression, is the key point of the bioprocess. Different strategies have been developed in Mut$^+$ phenotype [24, 32, 52, 102-105]. One of the most applied is a pre-programmed exponential feeding rate with an optimum methanol-glycerol ratio [38, 106], or similar strategy maintaining a residual methanol concentration between 1- 2 gl^{-1} [78]. The effect of different methanol-glycerol ratios at constant feeding rate has also been studied in the production of mouse α-amylase [107].

One important feature showed in these works is that, although the maximum specific growth rate of *P. pastoris* is around 0.2 h^{-1}, the optimum specific growth rate in Mut$^+$ phenotype is around 0.06 h^{-1}, too low compared with the maximum value. It seems that although glycerol is under limiting conditions high specific growth rate diminish the productivity of the bioprocess.

For this reason the use of different carbon sources other than glycerol may improve operational strategies on fed-batch cultures [99]. In contrast with glycerol, sorbitol accumulation during the induction phase does not affect the expression level of recombinant protein [108].

In shake flasks, inhibitory effect of sorbitol on cell growth appears at concentrations around 50 gl^{-1} [99]. Hence, control of residual sorbitol concentration during the induction phase is less critical than mixed feeds of glycerol and methanol. On the other hand less oxygen will be consumed during mixed substrate growth on sorbitol and methanol than using the combination glycerol and methanol or on methanol as sole carbon source [99]. However sorbitol has the disadvantage that the maximum specific growth rate is too low around 0.02 h^{-1} similar value that obtained in Mut^s phenotype growing on methanol as sole carbon source. Thus, time fermentation is long and sometimes the increase in the production not is reflected in the producitivity of the bioprocess.

Some different operational strategies have been implemented using sorbitol as co-substrate [99, 102, 106,109-114].

Arnau et al., [102,113] designed an operational strategy using a Mut^s phenotype comparing both co-substrates sorbitol and glycerol in the production of *Rhizopus oryzae* lipase [102,113]. The induction phase started with a preprogrammed exponential feeding rate of sorbitol or glycerol with the objective to maintain a constant specific growth rate under limiting substrate conditions. Methanol set-point was maintained using a predictive control algorithm coupled with a PI feedback control [89]. A set of different specific growth rates and methanol set-points were tested. When sorbitol was used as co-substrate the different specific growth rates tested did not have significance influence on specific production rate of the bioprocess, probably because the use of co-substrate improved the energetic state of the cells overcoming partially the unfolding protein response (UPR) and secretion problems observed in the production of this recombinant fungi lipase. The key parameter in terms of protein production was the methanol set-point selected. Optimal methanol concentration was 2 gl^{-1}, lower and higher concentrations diminished specific production rates. The product/biomass yield and the volumetric and specific productivity were 1.25-1.35 fold higher than using methanol as sole carbon source [113].

Irrespective of any economical reasons to use sorbitol or glycerol as co-substrate, one of the key advantages of using glycerol instead of sorbitol is its higher μ (0.2 h^{-1} versus 0.02 h^{-1}) and the subsequent potential increase in the productivity of the bioprocess. However, for Mut^s phenotype this potential advantage is ineffective, because when glycerol exceeds the μ_{max} of *P. pastoris* growing on methanol as a sole carbon source (around 0.014h^{-1}) a repression of *AOX* promoter is clearly observed, represented by a drastic decrease in methanol consumption rates. Additionally, when the relation μ_{Gly} per μ_{MeOH} was larger than 4, an important decrease of all productivity ROL parameters was observed. On the other hand, the presence of proteolytic activity detected when glycerol was used as co-substrate is another important drawback [102]. In conclusion sorbitol presented better results than glycerol as co-substrate in the heterologous production of *Rhizopus oryzae* lipase).

PAOX1 is strongly repressed by glucose at the transcription level. This is the cause that few authors present positive results using this substrate. Nevertheless, a real-time parameter-based controlled glucose feeding strategy has been developed successfully in the recombinant production of phytases [115], Mixtures of glucose and methanol has also been used in continuous cultures producing recombinant trypsinogen [116].

10. Alternative promoters

An important set of inducer promoters derived from genes which code for enzymes involved in the methanol metabolism are used as alternative promoters to the classical. *PAOX1*. A summary of the main alternative promoters is presented in table 2. Formaldehyde dehydrogenase promoter *PFLD1* inducible by methanol or methylamine [116], dihidroxyacetone synthase promoter *PDHAS* [101], and peroxisomal matrix protein gene promoter *PEX8* inducible by methanol or oleate [118] are some examples. Other inducer promoter is the isocitrate lyase 1 *PICL1*. This promoter is inducible with ethanol and repressed by glucose in the exponential phase, but not in the stationary phase [119].

Inducible promoters	Reference	Constitutive promoters	Reference
PAOX1	22	PGAP	121
PFLD1	116	PTEF1	122
PDHAS	101	PYPT1	123
PEX8	118	PPGK1	124
PICL1	119	PTHI1	120

Table 2. Summary of the main inducible and constitutive alternative promoters to *PAOX1*.

However, these alternative promoters have similar operational problems than *PAOX1*, especially when methanol is not substituted as inducer due to safety problems. This is the cause of a strong demand for alternative regulated promoters [120]. Between them, the constitutive glyceraldehydes-3-phosphate dehydrogenase promoter *PGAP* is the most common used [121]. Other constitutive promoters are the translation elongation factor 1-α promoter *PTEF1* [122], the promoter of YPT1, a GTPase involved in secretion [123] and the promoter of the 3-phosphoglycerate kinase *PPGK1*, from a glycolytic enzyme [124].

Stadlmayr *et al.,* [120] have identified 24 novel potential regulatory sequences from microarray data and tested their applicability to drive the expression of both, intracellular and secretory recombinant proteins with a broad range of expression levels. Although the production of model proteins not exceed the values obtained with the constitutive promoter *PGAP*, higher transcription levels at certain growth phases were detected with the translation elongation factor EF-1 promoter *PTEF1* and the promoter of a protein involved in the synthesis of the thiamine precursor *PTHI1*.

Between them only the inducer *PFLD1* and the constitutive *PGAP* have been applied for the routine production process, specially the last one.

The *FLD1* gene codes for an enzyme that plays an important role in the methanol catabolism as carbon source, as well as in the methylated amines metabolism as nitrogen source [125]. *PFLD1* is a strongly an independently induced either by methanol as carbon source or methylamine as nitrogen source [117]. Preliminary experiments to get an alternative carbon source to methanol showed that sorbitol, a carbon source that no repress the synthesis of methanol metabolism enzymes, also allows the induction of *PFLD1* by methylamine [126]. It suggests that the use of sorbitol as carbon source combined with methylamine as nitrogen source could be the basis for the development of methanol-free fed-batch fermentation. In fact, a methanol-free high cell density fed-batch strategy has been developed for the recombinant production of *Rhizopus oryzae* lipase. These fed-batch strategy has the same phases that a standard *PAOX1* promoter. GBP is similar but glycerol and ammonia as carbon and nitrogen sources are presented in a stoichiometric relation to achieve the exhaustion of both substrates at the end of the GBP. The TP consist in a sorbitol methylamine batch (SMBP) as a transition phase. The objective of the SMBP is the adaptation of the cells to the carbon and nitrogen sources used in the induction phase. Finally, the methylamine induction phase (MAIP) where a pre-programmed feeding rate strategy ensured a constant specific rate under sorbitol limiting conditions or maintaining a set-point of methanol at high specific growth rate have been implemented [127]. The result showed that the recombinant protein production is favored with the second strategy. When the performance of the bioprocess were compared to classical *PAOX1* promoter, the results were quite similar in terms of process productivity [63]. The production of this recombinant lipase under *PFLD1* triggers the unfolding protein response (UPR) detected at transcriptional levels [128].To overcome this problem two cell engineering strategies have been developed and applied successfully: the constitutive expression of the induced form of the *Saccharomyces cerevisiae* unfolded protein response transcriptional factor Hac1 and the deletion of the *GAS1*gene encoding a β-1,3 glucanosyltransglycosylase, GPI-anchored to the outlet leaflet of the plasma membrane, playing a key role in yeast cell wall assembly [129]. This is an example that how the co-expression of proteins or the deletion of genes affect to bioprocess engineering.

The great advantage of the constitutive GAP promoter is that the cloned heteroloogus protein will be expressed along with cell growth if the protein is not toxic for the cells [130]. The use of this promoter is more suitable for large-scale production because the hazard and cost associated with the storage and delivery of large volumes of methanol are eliminated [131], and also for the implementation of continuous cultures, continuous cultures practically not described using *PAOX1* [134]. Thus, the features of the GAP expression system may contribute significantly to the development of cost-effective methods for large-scale production of heterologous recombinants proteins [132-133]. The efficiency of *PGAP* compared with *PAOX1*depends generally of the protein expressed, although some times the better optimization of operational strategy can mask the results.

In general, the substrates used with this promoter are glucose or glycerol. The standard operational strategy is a batch phase using glycerol and a fed-batch phase in an open-loop control using glucose. The selection of the optimal sequence of both substrates is under studies. For instance, the production of rPEPT2 growing on glucose was approximately 2 and 8 times higher than in cells grown on glycerol and methanol [135].

When using this expression system, specific production rate increases asymptotically to a maximum value with increasing μ [68]. Maurer *et al.,* have developed a model to describe growth and product formation, optimizing the feeding profile of glucose limited fed batch cultures to increase volumetric productivity under aerobic conditions [68]. Under hypoxic conditions, where growth is controlled by carbon source limitation, while oxygen limitation was applied to modulate metabolism and heterologous protein productivity, an increase in the specific productivity has been observed. This strategy has additional benefits including lower aeration and lower final biomass concentration [73].

In conclusion PGAP is the most promise alternative to the classical PAOX1 promoter.

Acknowledgements

This work was supported by the project CTQ2010-15131 of the Spanish Ministry of Science and Innovation, 2009-SGR-281 and the Reference Network in Biotechnology (XRB) (Generalitat de Catalunya)

Author details

Francisco Valero*

Address all correspondence to: Francisco.Valero@uab.cat

Department of Chemical Engineering. Engineering School, Universitat Autònoma de Barcelona, Bellaterra, Barcelona, Spain

References

[1] Hitzeman RA, Hagie FF, Levine HL, Goeddel DW, Ammerer G, Hall BD. Expression of a human-gene for interferon in yeast, Nature 1981;293 717-722.

[2] Higgins DR, Cregg JM. Introduction to *Pichia pastoris*. In: Higgins DR, Cregg, JM (ed) *Pichia* protocols Methods in Molecular Biology, Vol. 103. Totowa, NJ: Humana Press Inc; 1998.

[3] Gellisen G, Hollenberg CP. Application of yeasts in gene expression studies: a comparison of *Saccharomyces cerevisiae*, *Hansenula polymorpha* and *Kluyveromyces* lactis – a review. Gene 1997; 90 87-97.

[4] Goffeau A, Barrell BG, Bussey H, Davis RW, Dujon B, Feldmann H, Galibert F, Hoheisel JD, Jacq C, Johnston M, Louis EJ, Mewes HW, Murakami Y, Philippsen P, Tettelin H, Oliver SG. Life with 6000 genes. Science 1996;274 546-567.

[5] Böer E, Steinborg G, Kunze G, Gellisen G. Yeast expression platform. Applied Microbiology and Biotechnology 2007;77 513-523.

[6] Giga-Hama Y, Thoda H, Takegawa K, Kumagai H. *Schizosaccharomyces pombe* minimum genome factory. Biotechnol Appl Biocehm 2007;46 147-155.

[7] Van Ooyen AJ, Dekker P, Huang M, Olsthoorn MM, Jacobs DI, Colussi PA, Taron CH. Heterologous protein production in the yeast *Kluyveromyces lactis*. FEMS Yeast Research 2006;6 381-392.

[8] Cereghino JL, Cregg JM. Heterologous protein expression in the methylotrophic yeast *Pichia pastoris*. FEMS Microbiology Reviews 2000;24 45-66.

[9] Sakai Y, Akiyama M, Kondoh H, Shibano Y, Kato N. High level secretion of fungal glucoamylase using the *Candida boindii* gene expression system. Biochimica Biophysica Acta 1996;1308 81-87.

[10]] Raymond CK, Bukowski T, Holderman SD, Ching AF, Vanaja E, Stamm MR. Development of the methylotrophic yeast *Pichia methanolica* for the expression of the 65 kilodalton isoform of human glutamate decarboxylase. Yeast 1998;14 11-23.

[11] Kang HA, Gellisen G. *Hansenula polymorpha*. In: Gellison G (ed) Production of recombinant proteins – novel microbial and eukaryotic expression systems. Weinheim: Wiley-VCH; 2005.

[12] Madzac C, Nicaud JM, Gaillardin C (2005). *Yarrowia lipolytica*. In: Gellison G (ed) Production of recombinant proteins – novel microbial and eukaryotic expression systems. Weinheim: Wiley-VCH; 2005.

[13] Böer E, Gellisen G, Kunze G (2005) *Arxula adeninivorans*. In: Gellison G (ed) Production of recombinant proteins – novel microbial and eukaryotic expression systems. Weinheim: Wiley-VCH; 2005.

[14] Madzac C, Gaillardin C, Beckerich JM. Heterologous protein expression and secretion in the non-conventional yeast *Yarrowia lipolytica*: a review. J Biotechnol 2004;109 63-81.

[15] Yin J, Li G, Ren X, Herrler G. Select what you need: A comparative evaluation of the advantages and limitations of frequently used expression systems for foreign proteins. Journal of Biotechnology 2007;127 335-347.

[16] Böer E, Steinborn G, Kunze G, Gellisen G. Yeast expression platforms. Applied Microbiology and Biotechnology 2007;77 513-523.

[17] Graf A, Dragosits M, Gasser B, Mattanovich D. Yeast systems biotechnology for the production of heterologous proteins. FEMS Yeast Research 2009;9 335-348.

[18] Idiris A, Tohda H, Kumagai H, Takegawa A. Engineering of protein secretion in yeast: strategies and impact on protein production. Applied Microbiology and Biotechnology 2010;86 403-417.

[19] Porro D, Gasser B, Fossati T, Maurer M, Branduardi P, Sauer M, Mattanovich D. Production of recombinant proteins and metabolites in yeasts. Applied Microbiology and Biotechnology 2011;89 939-948.

[20] Çelik E, Çalik P. Production of recombinant proteins by yeast cells. Biotechnology Advances 2012;30 1108-1118.

[21] Valero F., Heterologous expression system for lipases: A review. Methods in Molecular Biology 2012;861 161-178.

[22] Cregg JM, Vedvick TS, Raschke WV. Recent advances in the expression of foreign genes in *Pichia pastoris*. Bio/Technology 1993;11 905-910.

[23] Lin Cereghino GP, Cregg JM. Applications of yeast in biotechnology: protein production and genetic analysis. Current Opinions in Biotechnology 1999;10 422-427.

[24] Cos O, Ramón R, Montesinos JL, Valero F (2006) Operational strategies, monitoring and control of heterologous protein production in the methylotrophic yeast *Pichia pastoris* under different promoters: A review. Microbial Cell Factories 2006;5 1-20.

[25] Heterologous protein expressed in *Pichia pastoris*. http://www.kgi.edu/documents/faculty/James_Cregg/heterologous_proteins_expressed_in_pichia_pastoris.pdf. (accessed 3 September 2012)

[26] Yokohama S. Protein expression systems for structural genomics and proteomics. Current Opinion in Chemical Biology 2003;7 39-43.

[27] De Pourcq K, De Schutter K. Callewaert N. Engineering of glycosylation in yeast and other fungi: current state and perspectives. Applied Microbiology and Biotechnology 2010;87 1617-1631.

[28] Beck A, Cochet O, Wurch T. GlycoFi's technology to control the glycosylation of recombinant therapeutic proteins. Expert Opinion Drug Discovery 2010;5(1) 96-111.

[29] Cregg JM, Lin Cereghino J, Shi J, Higgins DR. Recombinant protein expression in *Pichia pastoris*. Molecular Biotechnology 2000;16 23-52.

[30] Daly R, Hearn MT., Expression of heterologous proteins in *Pichia pastoris*: a useful experimental tool in protein engineering and production. Journal of Molceular Rcognition 2005;18 119-138.

[31] Hohenblum H, Gasser B, Maurer M, Borth N, Mattanovich D. Effects of gene dosage, promoters, and substrates on unfolded protein stress of recombinant *Pichia pastoris*. Biotechnology and Bioengineering. 2004;85 367-375.

[32] Jahic M, Veide A, Charoenrat T, Teeri T, Enfors SO. Process technology for production of heterologous proteins with *Pichia pastoris*. Biotechnology Progress 2006;22 1465-1473.

[33] Cereghino GPL, Cereghino JL, Ilgen C, Cregg JM. Production of recombinant proteins in fermenter cultures of the yeast *Pichia pastoris*. Current Opinion in Biotechnology 2002;13 329-332.

[34] Harder W, Veenhuis M. Metabolism of one carbon compounds. In: Rose AH, Harrison JS (ed.) The Yeast. London: Academic press; 1989. P 289-316.

[35] Ellis SB. Isolation of alcohol oxidase and two other methanol regulatable genes from the yeast *Pichia pastoris*. Molecular Cell Biology 1985;5 1111-1121.

[36] Cregg, JM, Shen S, Johnson M, Waterham HR. Classical genetic manipulation. In: Higgins DR, Cregg, JM (ed) *Pichia* protocols Methods in Molecular Biology, Vol. 103. Totowa, NJ: Humana Press Inc; 1998 p17-26.

[37] Weast RC. Handbook of Chemistry and Physics. Boca Ratón (Florida): CRC Press Inc; 1980.

[38] Jungo C, Marison I, von Stockar U. Mixed feed of glycerol and methanol can improve the performance of *Pichia pastoris* cultures: A quantitative study based on concentration gradients in transient continuous cultures. Journal of Biotechnology 2007;128 824-837.

[39] Macauley-Patrick S, Fazenda ML, McNeil B, Harvey LM. Heterologous protein production using the *Pichia pastoris* expression system. Yeast 2005;22 249-270.

[40] Junker BH, Wang HY 2006. Bioprocess monitoring and computer control: key roots of the current PAT initiative. Biotechnology and Bioengineering 2006;95(2) 325-336.

[41] FDA. Guidance for industrial PAT-a Framework for innovative pharmaceutical manufacturing and quality assurance. Food and drug administration Rockville.

[42] Wechselberger P, Seifert A, Herwig C. PAT method to gather bioprocess parameters in real-time using simple input variables and first principle relationships. Chemical Engineering Science 2010;65 5734-5746.

[43] Teixeira AP, Duarte TM, Carrondo MJT, Alves PM. Synchronous fluorescence spectroscopy as a novel tool to enable PAT applications in Bioprocesses. Biotechnology and Bioengineering 2011;108(8) 1852-1861.

[44] Surribas A, Geissler D, Gierse A, Scheper T, Hitzmann B, Montesinos JL, Valero F. State variables monitoring by in situ multi-wavelength fluorescence spectroscopy in heterologous protein production by *Pichia pastoris*. Journal of Biotechnology 2006;124 412-419.

[45] Surribas A, Amigo JM, Coello J, Montesinos JL, Valero F, Maspoch S. Parallel factor analysis combined with PLS regression applied to the on-line monitoring of *Pichia pastoris* cultures. Analytical Bioanalytical Chemistry 2006;385 1281-1288.

[46] Amigo JM, Surribas A, Coello J, Montesinos JL, Maspoch S, Valero F. On-line parallel factor analysis.A step forward in the monitoring of bioprocesses in real time. Analytical Bioanalytical Chemometrics and Intelligent Laboratory Systems 2008;92 44-52.

[47] Barrigón JM, Ramón R, Rocha I, Valero F, Ferreira EC, Montesinos JL. State and specific growth estimation in heterologous protein production by *Pichia pastoris*. Aiche Journal 2012;58(10) 2966-2979.

[48] Cunha AE, Clemente JJ, Gomes R, Pinto F, Thomaz M, Miranda S, Pinto R, Moosmayer D, Donner P, Carrondo MJT: Methanol induction optimization for scFv antibody fragment production in *Pichia pastoris*: Biotechnology and Bioengineering 2004;86 458-467.

[49] Schenk J, Marison IW, von Stockar U. A simple method to monitor and control methanol feeding of *Pichia pastoris* fermentations using mid-IR spectroscopy. Journal of Biotechnology 2007; 128 344-353.

[50] Cos O, Serrano A, Montesinos JL, Ferrer P, Cregg JM, Valero F. Combined effect of the metanol utilization (Mut) phenotype and gene dosage on recombinant protein production in *Pichia pastoris* fed-batch cultures. Journal of Biotechnology 2005; 116 321-335.

[51] Potvin G, Ahmad A, Zhang Z. Bioprocess engineering aspects of heterologous protein production in *Pichia pastoris*: A review. Biochemical Engineering Journal 2012;64 91-105.

[52] Sreekrishna K. *Pichia*, Optimization of protein expression. In: Flickinger MC (ed.) Encyclopedia of Industrial Biotechnology: Bioprocess, Bioseparation, and Cell Technology. John Wiley & Sons Inc 2010 p 1-16.

[53] Ramón R, Feliu JX, Cos O, Monteinso JL, Berthet FX, Valero F. Improving the monitoring of methanol concentration during high cell density fermentation of *Pichia pastoris*. Biotechnology Letters 2004;26 144-1452.

[54] Surribas A, Cos O, Montesinso JL, Valero F. On-line monitoring of the methanol concentration in *Pichia pastoris* cultures producing an heterologous lipase by sequential injection analysis. Biotechnology Letters 2003;25 1795.1800.

[55] Gurramkonda C, Adnan A, Gäbel T, Lünsdorf H, Ross A, Nemani SK, Swaminathan S, Khanna N, Rinas U. Simple high-cell density fed-batch technique for high-level recombinant protein production with *Pichia pastoris*: application to intracellular production of hepatitis B surface antigen. Microbial Cell Factories 2009;8 13.

[56] Zhang Y, Yang B. In vivo optimizing of intracellular production of heterologous protein in *Pichia pastoris* by fluorescent scanning. Analytical Biochemistry 2006;357 232-239.

[57] Baker KN, Rendall MH, Patel A, Boyd P, Hoare M, Freedman RB, James DC. Rapid monitoring od recombinant protein products: a comparison of current technologies. Trends in Biotechnology 2002;20 149-156.

[58] Chaa HJ, Shin HS, Lim HJ, Cho HS, Dalal NN, Pham MQ, Bentley WE. Comparative production of human interleukin-2-fused with green fluorescent protein in several recombinant expression system. Biochemical Engineering Journal 2005;24 225-233.

[59] Cos O, Montesinos JL, Lafuente J, Sola C, Valero F. On-line monitoring of lipolytic activity by sequential injection analysis. Biotechnology Letters 2000;22 1783-1788.

[60] Cornelissen G, Leptien H, Pump D, Scheffler U, Sowa E, Radeke HH, Luttmann R. Integrated bioprocess development for production of recombinant proteins in high cell density cultivations with *Pichia pastoris*. CAB8 Computer Applications in Biotecnology 2001.

[61] Invitrogen corporation http://www.invitrogen.com (accessed 3 September 2012)

[62] Brierley RA, Bussineau C, Kosson R, Melton A, Siegel RS. Fermentation development of recombinant *Pichia pastoris* expressing heterologous gene: bovine lysozyme. Annal New York Academic of Science 1990;589: 350-362.

[63] Cos O, Resina D, Ferrer P, Montesinos JL, Valero F. Heterologous protein production of *Rhizopus oryzae* lipase in *Pichia pastoris* using the alcohol oxidase an formaldehyde dehydrogenase promoters in batch and fed-batch cultures. Biochemical Engineering Journal 2005;26 86-94.

[64] Jahic M, Rotticci-Mulder JC, Martinelle M, Hult K, Enfors SO. Modeling of growth and energy metabolism of *Pichia pastoris* producing a fusion protein. Bioprocess Biosystem Engineering 2002;24 385-393.

[65] Chiruvolu V, Eskridge K, Cregg J, Meagher M. Effects on glycerol concentration and pH on growth of recombinant *Pichia pastoris* yeast. Applied Biochemistry and Biotechnology 1998;75 163-173.

[66] Zhang W, Inan M, Meagher MM. Fermentation strategies for recombinant protein expression in the methylotrophic yeast *Pichia pastoris*. Biotechnology Bioprocess Engineering 2000;5 275-287.

[67] Maurer M, Kühleitner M, Gasser B, Mattanovich D. Versatile modeling and optimization of fed-batch processes for the production of secreted heterologous proteins with *Pichia pastoris*. Microbial Cell Factories 2006;5(37) 1-10.

[68] Sing S, Gras A, Vandal CF, Ruprecht J, Rana R, Martinez M, Strange PG, Wagner R, Byrne B. large-scale functional expression of WT and truncated human adenosine A2A in *Pichia pastoris* bioreactor cultures. Microbial Cell Factories 2008:7 28.

[69] Khatri NK, Hoffmann F. Impact of methanol concentration on secreted protein production in oxygen-limited cultures or recombinant *Pichia pastoris*. Biotechnology and Bioengineering 2005;93 871-879.

[70] Khatri NK, Hoffmann F. Oxygen-limited control of methanol uptake for improved production of a single-chain antibody fragment with recombinant *Pichia pastoris*. Applied Microbiology and Biotechnology 2006;72 492-498.

[71] Charoenrat T, Ketudat-Cairns M, Sthendahl-Andersen H, Jahic M, Enfors SO. Oxygen-limited fed-batch process: an alternative control for *Pichia pastoris* recombinant protein processes. Bioprocess and Biosystem Engineering 2005;27 399-406.

[72] Baumann K, Maurer M, Dragosits M, Cos O, Ferrer P, Mattanovich D. Hypoxic fed-batch cultivation of *Pichia pastoris* increases specific and volumetric productivity of recombinant proteins. Biotechnology and Bioengineering 2008;100 177-183.

[73] Potgieter TI, Kersey SD, Mallem MR, Nylen AC, d'Anjou M. Antibody expression ki-
      netics in glycoengineered *Pichia pastoris*. Biotechnology and Bioengineering 2010;
      106(6) 918-927.

[74] Berdichevsky M, d'Anjou M, Mallem MR, Shaikh SS, Potgieter TI. Improved produc-
      tion of monoclonal antibodies through oxygen-limited cultivation of glycoengineered
      yeast. Journal of Biotechnology 2011;155 217-224.

[75] Ye J, Ly J, Watts K, Hsu A, Walker A, McLaughlin K, Berdichevsky M, Prinz B, Ker-
      sey DS, d'Anjou M, Pollard D, Potgieter T. Optimization of a glycoengineered *Pichia
      pastoris* cultivation process for commercial antibody production. Biotechnology Prog-
      ress 2011;27(6) 1744-1750.

[76] Chung, JD. Design of metabolic feed controllers: Application to high-density fermen-
      tations of *Pichia pastoris*. Biotechnology and Bioengineering 2000;68 298-307.

[77] D'Anjou MC, Daugulis AJ. A rational approach to improving productivity in re-
      combinant *Pichia pastoris* fermentation. Botechnology and Bioengineering 2000;72
      1-11.

[78] Hu XQ, Chu J, Zhang Z, Zhang SL, Zhyang YP, Wang YH, Guo MJ, Chen HX, Yuan
      ZY. Effects of different glycerol feeding strategies on S-adenosyl-l-methionine bio-
      synthesis by PGAP-driven *Pichia pastoris* overexpressing methionine adenosyltrans-
      ferase. Journal of Biotechnology 2008;137 44-49.

[79] Oliveira R, Clemente JJ, Cunha AE, Carrondo MJT. Adaptive disolved oxygen con-
      trol through the glycerol feeding in a recombinant *Pichia pastoris* cultivation in condi-
      tions of oxygen transfer limitation. Journal of Biotechnology 2005;116 35-50.

[80] Zhang W, Bevins MA, Plantz BA, Smith LA, Meagher MM. Modeling *Pichia pastoris*
      growth on methanol and optimizing the production of a recombinant protein, the
      heavy-chain fragment C of *Botulinum* neurotoxin serotype A. Biotechnology and Bio-
      engineering 2000;70 1-8.

[81] Ren HT, Yuan J, Bellgardt KH. Macrokinetic model for methylotrophic *Pichia pastoris*
      based on stoichiometric balance. Journal of Biotechnology 2003;106 53-68.

[82] Sinha J, Plantz BA, Zhang W, Gouthro M, Schlegel VL, Liu CP, Meagher MM. Im-
      proved production of recombinant ovine interferon-τ by Mut+ strain of *Pichia pastoris*
      using an optimized methanol feed profile. Biotechnology Progress 2003;19 794-802.

[83] Trinh LB, Phue JN, Shiloah J. Effect of methanol feeding strategies on production and
      yield of recombinant mouse endostatin from *Pichia pastoris*. Biotechnology and Bio-
      engineering 2003;82 438-444.

[84] Bawa Z, Darby RAJ. Optimising *Pichia pastoris* induction. Methods in Molecular Biol-
      ogy 2012;866 181-190.

[85] Katakura Y, Zhang WH, Zhuang GQ, Omasa T, Kishimoto M, Goto W, Suga KI. Ef-
      fect of methanol concentration on the production of human beta(2)-glycoprotein I do-

main V by a recombinant *Pichia pastoris*: A simple system for the control of methanol concentration using a semiconductor gas sensor. Journal of Fermentation and Bioengineering 1998;86 482-487.

[86] Guarna MM, Lesnicki GJ, Tam BM, Robinson J, Radziminski CZ, Hasenwinkle D, Boraston A, Jervis E, Macgillivray RTA, Turner RFB, Kilburn DG. On line monitoring and control of methanol concentration in shake-flasks cultures of *Pichia pastoris*. Biotechnology and Bioengineering 1997; 56 279-286.

[87] Zhang WH, Smith LA, Plantz BA, Siegel Vl, Meagher MM. Design of methanol feed control in *Pichia pastoris* fermentations based upon a growth model. Biotechnology Progress 2002;18 1392-1399.

[88] Cos O, Ramón R, Montesinos JL, Valero F. A simple model-based control for *Pichia pastoris* allowas a more efficient heterologous protein production bioprocess. Biotechnology and Bioengineering 2006;95(1) 145-1154.

[89] Curvers S, Brixius P, Klauser T, Thömmes J, Weuster-Botz D, Takors R, Wandrey C. Human chymotrypsinogen B production with *Pichia pastoris* by integrated development of fermentation and downstream processing. Part I. Fermentation. Biotechnology Progress 2001;17 495-502.

[90] Surribas A, Stahn R, Montesinos JL, Enfors SO, Valero F, Jahic M. Production of a *Rhizopus oryzae* lipase from *Pichia pastoris* using alternative operational strategies. Journal of Biotechnology 2007; 130 291-299.

[91] Jahic M, Wallberg F, Bollok M, García P, Enfors SO. Temperature limited fed-batch technique for control of proteolysis in *Pichia pastoris* bioreactor cultures. Microbial Cell Factories 2003;2 1-6.

[92] Siren N, Weegar J, Dahlbacka J, Kalkkinen N, Fagervik K, Leisola M, von Weymarn N. Production of recombinant HIV-1 Nef (negative factor) protein using *Pichia pastoris* and a low-temperature fed-batch strategy. Biotechnology and Applied Biochemistry 2006; 44 151-158.

[93] Yang M, Johnson SC, Murthy PN. Enhancement of alkaline phytase production in *Pichia pastoris*: Influence of gene dosage, sequence optimization and expression temperature. Protein Expression and Purification 2012;84(2) 247-254.

[94] Dragosits M, Frascotti G, Bernard-Granger L, Vazquez F, Giuliani M, Baumann K, Rodriguez-Carmona E, Tokkanen J, Parrilli E, Wiebe MG, Kunert R, Maurer M, Gasser B, Sauer M, Branduardi P, Pakula T, Saloheimo M, Penttila M, Ferrer P, Tutino ML, Villaverde A, Porro D, Mattanovich D. Influence of Growth Temperature on the Production of Antibody Fab Fragments in Different Microbes: A Host Comparative Analysis. Biotechnology Progress 2011;27(1) 38-46.

[95] Dragosits M, Stadlmann J, Albiol J, Baumann, K, Maurer M, Gasser B, Sauer M, Altmann F, Ferrer P, Mattanovich D. The Effect of Temperature on the Proteome of Recombinant *Pichia pastoris*. Journal of proteome research 2009;8(3) 1380-1392.

[96] Trentmann O, Khatri NK, Hoffmann F: reduced oxygen supply increases process stability and product yield with recombinant *Pichia pastoris*. Biotechnology Progress 2004;20 1766-1775.

[97] Narendar KK, Hoffmann F. Impact of methanol concentration on secreted protein production in oxygen-limited cultures of recombinant *Pichia pastoris*. Biotechnology and Bioengineering 2006;93 871-879.

[98] Jungo C, Schenk J, Pasquier M, Marison IW, von Stockar U. cultures: A quantitative analysis of the benefits of mixed feeds of sorbitol and methanol for the production of recombinant avidin with *Pichia pastoris*. Journal of Biotechnology 2007;131 57-66.

[99] Von Stockar U, Gustafsson L, Larsson C, Marison I, Tissot P, Gnaiger E. Thermodynamic considerations in constructing energy balances for cellular growth, Biochemistry and Biophysic Acta 1993;1183 221-240.

[100] Tschopp JF, Brust PF, Cregg JM, Stillman CA, Gingeras TR. Expression of the lacZ gene from two methanol-regulated promoters in *Pichia pastoris*, Nucleic Acids Res. 15 (1987) 3859-3876.

[101] Arnau C, Casas C, Valero F. The effect of lycerol mixed substrate on the heterologous production of *Rhizopus oryzae* lipase in *Pichia pastoris* system. Biochemical Engineering Journal 2011;57 30-37.

[102] Gao M-J, Zheng Z-Y, Wu J-R, Dong S-J, Jin Z-L, Zhan X-B, Lin C-C. Improvement of specific growth rate of *Pichia pastoris* for effective porcine interferon-α production with an on-line model based glycerol feeding strategy. Applied Microbiology and Biotechnology 2012;93-1437-1445.

[103] Zalaid D, Dietzsch C, Herwig C, Spadiut O. A dynamic fed batch strategy for a *Pichia pastoris* mixed feed system to increase process understanding. Biotechnology Progress 2012;28(3) 878-886.

[104] Huang , Yang P, Luo H, Tang H, Shao N, Yuan T, Wang Y, Bai Y, Yao B. High-level expression of a truncated 1,3-1,4-β-D-glucanase from *Fibrobacter succinogenes* in *Pichia pastoris* by optimization of codons and fermentation. Applied Microbiology and Biotechnology 2008;103 78-95.

[105] Boze H, Laborde C, Chemardin P, Richard F, Venturin C, Combarnous Y, Moulin G. High-level secretory production of recombinant porcine follicle-stimulating hormone by *Pichia pastoris*, Process Biochemistry. 2001;36 907-913.

[106] Choi DB, Park EY. Enhanced production of mouse α-amilase by feeding combined nitrogen and carbon sources in fed-batch culture of recombinant *Pichia pastoris*. Process Biochemistry 2006;41 390-397.

[107] Resina D, Cos O, Ferrer P, Valero F. Developing high cell density fed-batch cultiva-
      tion strategies for heterologous protein production in *Pichia pastoris* using the nitro-
      gen source-regulated *FLD1* promoter. Biotechnology and Bioengineering
      2005;91:760–767.

[108] Sreekrishna K, Brankamp RG, Kropp KE, Blankenship DT, Tsay JT, Smith PL,
      Wierschke JD, Subramaniam A, Birkenberger LA. Strategies for optimal synthesis
      and secretion of heterologous proteins in the methylotrophic yeast *Pichia pastoris*.
      Gene 1997;190 55-62.

[109] Inan M, Meagher MM. Non-represing carbon sources for alcohol oxidase *(AOX1)*
      promoter of *Pichia pastoris*. Journal of Bioscience and Bioengineering 2001;92 585-589.

[110] Thorpe ED, D'Anjou MC, Daugulis A. Sorbitol as a non-represing carbon source for
      fed-batch fermentation of recombinant *Pichia pastoris*. Biotechnology Letters 1999;21
      669-672.

[111] Xie JL, Zhou QW, Pen D, Gan RB, Qin Y. Use of different carbon sources in cultiva-
      tion of recombinant *Pichia pastoris* for angiostatin production. Enzyme and Microbial
      Technology 2005;36 210-216

[112] Arnau C, Ramón R, Casas C, Valero F. Optimization of the heterologous production
      of *Rhizopus oryzae* lipase in *Pichia pastoris* system using mixed substrates on control-
      led fed-batch bioprocess. Enzyme and Microbial Technology 2010;46 494-500.

[113] Çelik E, Çalik P, Oliver SG. Fed-batch methanol feeding strategy for recombinant
      protein production by *Pichia pastoris* in the presence of co-substrate sorbitol. Yeast
      2009;26 473-484.

[114] Hang H, Ye XY, Guo M, Chu J, Zhuang Y, Zhang M, Zhang S. A simple fermentation
      strategy for high-level production of recombinant phytase by *Pichia pastoris* using
      glucose as the growth substrate. Enzyme and Microbial Technology 2009;44 185-188.

[115] Paulova L, Hyka P, Branska B, Melzoch K, Kovar K. Use of a mixture of glucose and
      methanol as substrates for the production of recombinant trypsinogen in continuous
      cultures with *Pichia pastoris* Mut+. Journal of Biotechnology 2012;157 180-188.

[116] Shen S, Sulter G, Jeffries TW, Cregg JM. A strong nitrogen source-regulated promoter
      for controlled expression of foreign genes in the yeast *Pichia pastoris*. Gene
      1998;216(1) 93-102.

[117] Liu H, Tan X, Rissell KA, Veenhuis M, Cregg JM. *ER3*, a gene required for peroxi-
      some biogenesis in *Pichia pastoris*, encodes a peroxisomal membrane protein involved
      in protein import. Journal Biological Chemistry 1995;270 10940-10951.

[118] Menendez J, Valdes I, Cabrera N. The ICL1 gene of *Pichia pastoris*, transcriptional reg-
      ulation and use of its promoter. Yeast 2003;20(13) 1097-1108.

[119] Menendez J, Valdes I, Cabrera N. The ICL1 gene of *Pichia pastoris*, transcriptional regulation and use of its promoter. Yeast 2003;20(13) 1097-1108.

[120] Stadlmayr G, Mecklenbräuer A, Rothmüller M, Maurer M, sauer M, Mattanovich D, Gasser B. Identification and characterization of novel *Pichia pastoris* promoters for heterologous protein production. Journal of Biotechnology 2010;150 519-529.

[121] Waterham HR, Digan ME, Koutz PJ, Lair SV, Cregg JM. Isolation of the *Pichia pastoris* glyceraldehydes-3-phosphate dehydrogenase gene and regulation nd use of its promoter. Gene 1997;186(1) 37-44.

[122] Ahn J, Hong J, Lee H, Park M, Lee E, Kim C, Choi E, Jung J. Translation elongation factor 1-alpha gene from *Pichia pastoris*: molecular cloning, sequence, and use of its promoters. Applied Microbiology and Biotechnology 2007; 74(3) 601-608.

[123] Sears I, O'Connor J, Rossanese O, Glick B. A versatile set of vectors for constitutive and regulated gene expression in *Pichia pastoris*. Yeast 1998;14(8) 783-790.

[124] De Almeida JR, de Moraes LM, Torres FA. Molecular characterization of the 3-phosphoglycerate kinase gene (PGK1) from the methylotrophic yeast *Pichia pastoris*. Yeast 2005;22(9) 725-737.

[125] Harder W, Veenhuis M. Metabolism of one carbon compounds. In: Rose AH, Harrison JS (ed.). The Yeasts. London: Academic Press; 1989. 289-326.

[126] Thorpe ED, D'Anjou MC, Daugulis AJ. Sorbitol as a non repressing carbon-source for fed-batch fermentation of recombinant *Pichia pastoris*. Biotechnology Letters 1999;21: 669-672.

[127] Resina D, Cos O, Ferrer P, Valero F. Developing high cell density fed-batch cultivation strategies for heterologous protein production in *Pichia pastoris* using the nitrogen source-regulated *FLD1* promoter. Biotechnology and Bioengineering 2005; 91(6) 760-767.

[128] Resina D, Bollock M, Khatri NK, Valero F, Neubauer P, Ferrer P. Transcriptional response of *Pichia pastoris* in fed-batch cultivations to *Rhizopus oryzae* lipase production reveals UPR induction. Microbial Cell Factories 2007;6(21) 1-11.

[129] Resina D, Maurer M, Cos O, Arnau C, Carnicer M, Marx H, Gasser B, Valero F, Mattanovich D, Ferrer P. Engineering of bottlenecks in *Rhizopus oryzae* lipase production in *Pichia pastoris* using the nitrogen source-regulated *FLD1* promoter. New Biotechnology 2009;25(6) 396.403.

[130] Boer H, Teeri TT, Koivula A. Characterization of *Trichoderma reesei* cellobiohydrolase Cel7A secreted from *Pichia pastoris* using two different promoters. Biotechnology and Bioengineering 2000;69 486-494.

[131] Goodrick JC, Xu M, Finnegan R, Schilling BM, Schiavi S, Hoppe H, Wan NC. High-level expression ans stabilization of recombinant human chitinase produced in a con-

tinuous constitutive *Pichia pastoris* expression system. Biotechnology and Bioengineering 2004;31 330-334.

[132] Wu JM, Lin JC, Chieng LL, Lee CK, Hsu TA. Combined used of GAP and AOX1 promoter to enhance the expression of human granulocyte-macrophage colony-stimulating factor in *Pichia pastoris*. Enzyme and Microbial Technology 2003;33 453-459.

[133] Delroise JM, Dannau M, Gilsoul JJ, El Mejdoub T, Destain J, Portetelle D, Thonart P, Haubruge E, Vandelbol M. Expression of a synthetic gene encoding a tribolium castaneum carboxylesterase in *Pichia pastoris*. Protein Expression and Purification 2005; 42:286-294.

[134] Zhang A-L, Luo J-X, Zhang T-Y, Pan Y-W, Tan Y-H, Fu C-Y, Tu F-z. Recent advances on the GAP promoter derived expression system of *Pichia pastoris*. Molecular Biology Reports 2009;36 1611-1619.

[135] Dóring F, Klapper M, Theis S, Daniel H. Use of the glyceraldehydes-3-phosphate dehdrogenase promoter for production of functional mammalian membrane transport proteins in the yeast *Pichia pastoris*. Biochemistry Biophysics Research Communication 1998;250. 531-535.

Application

Applications of the *In Vitro* Virus (IVV) Method for Various Protein Functional Analyses

Noriko Tabata, Kenichi Horisawa and
Hiroshi Yanagawa

Additional information is available at the end of the chapter

1. Introduction

The complete decoding of the human genome in 2003 raised expectations for targeted drug design and personalized medical care based on genetic information and also for gene therapy of disease. To reach these goals, it is necessary to develop new technologies capable of identifying drug-target genes from enormous volumes of genome information, validating the biological functions, and comprehensively analyzing protein functions and interactions that have hitherto been studied individually. We first developed an mRNA display, termed the *in vitro* virus (IVV) method [1-3], and a C-terminal protein labeling method [4,5]that provided the required technology (Fig. 1).

The IVV method, originally developed for evolutionary protein engineering based on *in vitro* translation systems, was subsequently applied for the analysis of various protein interactions. In the IVV method, the genotype molecule (mRNA) is linked to the phenotype molecule (protein) through puromycin in a cell-free translation system [3]. We have developed this method into a stable, efficient, and high-throughput technique that allows simple selection without any requirement for post-translational work [3], unlike previous systems. An additional technological advance is the elimination of the need to express and purify bait proteins for downstream protein-protein interaction studies. Our totally *in vitro* cell-free co translation system provides a simpler solution that is suitable for high-throughput, genome-wide analysis, as baits are synthesized within each reaction. Moreover, as cotranslation of bait and prey proteins is favorable for the formation of multi-protein complexes, this approach offers a better chance to obtain a more comprehensive data set, including both direct and indirect interactions, in a single experiment. Here, we provide an overview of this system and discuss its advantages for analyses of various protein interactions including protein-protein, DNA(RNA)-protein, peptide-protein, drug-protein, and antigen-antibody interactions.

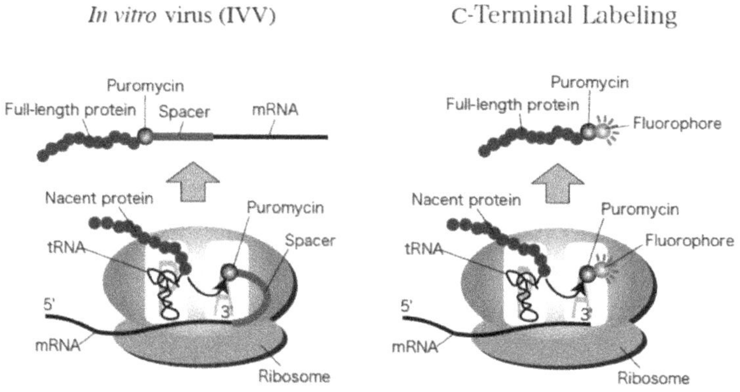

In vitro virus (IVV)   C-Terminal Labeling

Figure 1. *In vitro* virus(IVV) formation and C-terminal labeling on the ribosome. Puromycin at the 3'-terminal end of a spacer ligated to mRNA can enter the ribosomal A-site to bind covalently to the C-terminal end of the encoded full-length protein in the ribosomal P-site. The same property of puromycin to develop a highly specific labeling system for proteins was also utilized, in which puromycin derivatives bearing a fluorescein moiety are used to label the C-terminal end of a full-length protein.

The antibiotic puromycin [6], which is an analogue of the 3' end of TyrtRNATyr [7], acts in both prokaryotes and eukaryotes [8] as an inhibitor of peptidyl transferase [9]. It has two modes of inhibitory action. The first is by acting as an acceptor substrate that attacks peptidyl-tRNA (donor substrate) in the P site to form a nascent peptide [10]. The second is by competing with aminoacyl-tRNA for binding to the A' site, which is the binding site of the 3' end of aminoacyl-tRNA within the peptidyl transferase active site [11]. It has been reported that the polypeptides released by puromycin are not full-length proteins [9]. Similarly, it has been shown that growing peptide chains on ribosomes are transferred to the α-amino group of puromycin, which interrupts the normal reaction of peptide bond formation [10]. Therefore, these conventional studies suggested that puromycin is a non-specific inhibitor of protein synthesis as a result of competition with aminoacyl-tRNA. However, since most of the studies on puromycin were performed at relatively high concentrations, the behavior of puromycin at lower concentrations was still an open question at that time. It had been reported that full-length protein that fails to be released from ribosomes at the final stage of protein synthesis, requires treatment with puromycin or RFs to be released [12]. These results led us to hypothesize that puromycin at very low concentrations, which would not effectively compete with aminoacyl-tRNA [2], might act as a noninhibitor and bond specifically to full-length protein at the stop codon. Indeed, we confirmed that puromycin and its derivatives bond only to full-length protein at very low concentrations (such as 0.04 μM), where they are "non-inhibitors" of protein synthesis, by using ^{32}P-labeled rCpPuro (2'-ribocytidylyl-(3'→5')- puromycin) in an *E. coli* S30 extract cell-free translation system [2]. Our results provided the first evidence that specific bonding of puromycin to the full-length protein occurs at the stop codon during the process of termination of protein synthesis at very low concentration. In other words, puromycin at sufficiently low concentrations only has the opportunity to be bonded to proteins at a stop codon, where it does not need to compete with aminoacyl-tRNA. Since termination is a relatively slow step involving a translational pause in eukaryotes [13] and *E. coli* [14], it is possible that puromycin even at very low concentrations can bind to the A' site, and compete with RFs to release the full-length protein from ribosomes. Accordingly, under these conditions, puromycin can be incorporated specifically at the C-terminus of the full-length protein. This concept is the basis of so-called puromycin technology [2,3]. The combination of this puromycin technology with *in vitro* translation has yielded novel methods, such as IVV and C-terminal labeling techniques, for protein selection and screening [15,16], fluorescence labeling [2,5], and affinity purification (pull-down assay), as well as protein chips for proteomics [5,17], and superior methods for evolutionary protein engineering.

2. Application of the IVV method for analyzing various protein interactions

The IVV method is applicable for analyzing protein–protein [15,16,18,19] DNA(RNA)–protein [20-22], peptide–protein [23-27], drug–protein [28, 29] and antigen–antibody [30,31] interactions. As a result, we are able to explore protein complexes, transcription factors, RNA-binding proteins, bioactive peptides, drug-target proteins, and antibodies (antibodies will be discussed later) (Fig. 2). library and conjugation with a polyethylene glycol (PEG) spacer with puromycin, cell-free translation to form IVV and interaction of bait attached to beads and prey IVV library to form complexes, IVV selection with bait, and RT-PCR to amplify mRNA tags. Selection rounds are repeated until sufficient enrichment is obtained, followed by cloning and sequencing, and protein sequences are decoded.

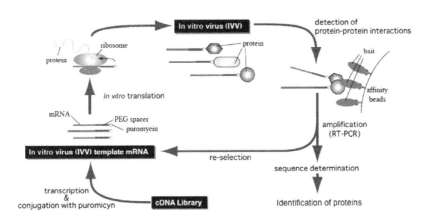

Figure 2. IVV selection based on cell-free co translation. The IVV selection procedure is composed of transcription from a cDNA

2.1. DNA-protein interaction

The specific interactions between cis-regulatory DNA elements and transcription factors are critical components of transcriptional regulatory networks [32,33]. The whole genome and complete cDNA sequences contain large numbers of transcription factors and their binding DNA sequences, and thus comprehensive analysis of DNA-transcription factor interactions is expected to provide a deep understanding of the mechanisms of cell proliferation, developmental processes in tissue morphogenesis and disease. Currently, combined use of chromatin immunoprecipitation (ChIP) assay with DNA microarrays (ChIP-chip)[34] has become the most widely used high-throughput method for discovering cis-regulatory DNA elements for a transcription factor. In contrast, development of high-throughput methods for discovering transcription factors for a cis-regulatory DNA element remains at an early stage. Although the yeast one-hybrid method [35] and phage display [36] are attractive candidates, these methods are not easily scalable because of the use of living cells. In addition, as over-expression of transcription factors often affects cellular metabolism, such transcription factors are difficult to screen. In order to circumvent these difficulties, we focused on a totally *in vitro* mRNA display technology such as IVV method [1-3,15,16] for the discovery of DNA–protein interactions.

Comprehensive analysis of DNA–protein interactions is important for mapping transcriptional regulatory networks at a genome-wide level. We employed the IVV method for *in vitro* selection of DNA-binding protein heterodimeric complexes [20]. Under improved selection conditions using a TPA-responsive element (TRE) as a bait DNA, known interactors c-fos and c-jun were simultaneously enriched about 100-fold from a model library (a 1:1:20 000 mixture of c-fos, c-jun and gst genes) after one round of selection. Furthermore, almost all of the AP-1 family genes, including c-jun, c-fos, junD, junB, atf2 and b-atf, were successfully selected from an IVV library constructed from a mouse brain poly A^+ RNA after six rounds of selection. These results indicate that the IVV selection system can identify a variety of DNA-binding protein complexes in a single experiment. Since almost all transcription factors form hetero-oligomeric complexes to bind with their target DNA, this method should be most useful to search for DNA-binding transcription factor complexes.

2.2. Peptide–protein interaction

Peptides are powerful tools for disrupting protein-protein interactions because the large interacting surfaces and the high specificity of these peptides lead to fewer adverse side effects when they are used as pharmaceutical agents [37]. As previously reported, several peptides that inhibit the MDM2-p53 interaction have been identified from randomized peptide libraries using phage display [38]. Hu et al. identified a 12-amino-acid (aa) peptide (LTFEHYWAQLTS), DI, that could inhibit not only the MDM2-p53 interaction, but also the MDMX-p53 interaction more effectively than Nutlin-3, a small molecular inhibitor of the MDM2-p53 interaction [39]. An MDM2 homologue, MDMX is highly expressed in tumors, and it binds to and negatively regulates p53 [40]. Furthermore, DI expressed with recombinant adenovirus as a thioredoxin-fused protein could activate the p53 pathway both *in vitro* and *in vivo*. However, DI was not sufficiently optimized because it was selected by phage display from a 12-mer random library (4.161,015 possible members) with a size of $\sim 10^8$ that did not cover all of the possible sequences.

To overcome this problem, we performed *in vitro* selection of MDM2-binding peptides from random peptide libraries using the IVV method [1-3,15]. This system based on cell-free translation is a potent method for screening large peptide libraries (10^{13} unique members) and is able to cover all of the possible sequences in a 10-mer random library. We applied the IVV method to identify a highly optimized peptide that could disrupt the MDM2-p53 complex from a random library containing all of the possible sequences by dividing the selection process into two stages. We also verified that a selected peptide could inhibit the MDM2-p53 interaction in living cells and block tumor cell growth.

We identified an optimal peptide named MIP that inhibited the MDM2-p53 and MDMX-p53 interactions 29- and 13-fold more effectively than DI, respectively (Fig. 3) [23]. Adenovirus-mediated expression of MIP fused to the thioredoxin scaffold protein in living cells caused stabilization of p53 through its interaction with MDM2, resulting in activation of the p53 pathway. Furthermore, expression of MIP also inhibited tumor cell proliferation in a p53-dependent manner more potently than did DI. These results show that two-stage, the peptide selection by IVV method [1-3,15] is useful for the rapid identification of potent peptides that target oncoproteins.

Bcl-X_L, an antiapoptotic member of the Bcl-2 family, is a mitochondrial protein that inhibits activation of Bax and Bak, which commit the cell to apoptosis, and it therefore represents a potential target for drug discovery. Peptides have potential as therapeutic molecules because they can be designed to engage a larger portion of the target protein with higher specificity. We selected 16-mer peptides that interact with Bcl-X_L from random and degenerate peptide libraries using the IVV method [24]. The selected peptides have sequence similarity with the Bcl-2 family BH3 domains, and one of them has higher affinity (IC_{50} = 0.9 μM) than Bak BH3 (IC_{50} = 11.8 μM) for Bcl-X_L *in vitro*. We also found that GFP fusions of the selected peptides specifically interact with Bcl-X_L, localize in mitochondria, and induce cell death. Further, a chimeric molecule, in which the BH3 domain of Bak protein was replaced with a selected peptide, retained the ability to bind specifically to Bcl-X_L. These results demonstrate that this selected peptide specifically antagonizes the function of Bcl-X_L and overcomes the effects of Bcl-X_L in intact cells. Thus, the IVV method is a powerful technique to identify peptide inhibitors with high affinity and specificity for disease-related proteins.

The importin α/β pathway mediates nuclear import of proteins containing the classical nuclear localization signals (NLSs). Although the consensus sequences of the classical NLSs have been defined, there are still many NLSs that do not match the consensus rule and many nonfunctional sequences that match the consensus. We identified six different NLS classes that specifically bind to distinct binding pockets of importin α. By screening of random peptide libraries using the IVV method, we selected peptides bound by importin α and identified six classes of NLSs, including three novel classes (Table 1). Two noncanonical classes (class 3 and class 4) specifically bound to the minor binding pocket of importinα, whereas the classical monopartite NLSs (class 1 and class 2) bound to the major binding pocket. Using a newly developed universal green fluorescent protein expression system, we found that these NLS classes, including plant-specific class 5 NLSs and bipartite NLSs, fundamentally require regions outside the core basic residues for their activity and have specific residues or patterns that confer distinct activities between yeast, plants, and mammals. Furthermore, amino acid replacement analyses revealed that the consensus basic patterns of the classical NLSs are not essential for activity, and more unconventional patterns, including redox-sensitive NLSs, were generated. These results explain the causes of NLS diversity. The defined consensus patterns and properties of importin α-dependent NLSs provide useful information for identifying NLSs [25-27].

Clone No.	Peptide sequence	Frequency
X12-1 (MIP)	PRFWEYWLRLME	10
X12-2	KSFQQYWQELML	9
X12-3	KTFEEYWLMLMS	4
X12-4	PSFWEHWVELML	4
X12-5	KRFQDYWSELML	3
p53$_{17-28}$	ETFSDLWKLLPE	-
	19 23 26	

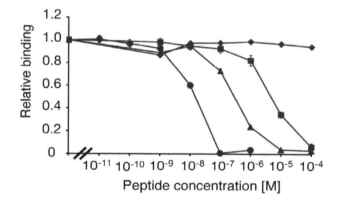

Figure 3. A multiple-sequence alignment of the selected peptides(upper) and inhibition of MDM2-p53 interactions by synthetic peptides(lower). Peptide sequences selected from randomized mRNA displayed peptide libraries were aligned using the ClustalW program. 12-mer peptides after five rounds of selection. Fixed amino acid residues are shown in bold. Values below the displayed p53$_{17-28}$ sequence indicate the position of amino acids in the p53 protein. MDM2 was generated by an *in vitro* transcription reaction, bound to His6-p53 immobilized on copper-coated plates in the presence of various concentrations of synthetic MIP(circle), DI (triangle), 3A (diamond) or p53 (square) peptides and quantified by ELISA.

NLS class	Consensus sequence[a]
Class 1	KR(K/R)R, K(K/R)RK
Class 2	(P/R)XXKR($^\wedge$DE)(K/R)
Class 3	KRX(W/F/Y)XXAF
Class 4	(R/P)XXKR(K/R)($^\wedge$DE)
Class 5	LGKR(K/R)(W/F/Y)
Bipartite	KRX_{10-12}K(KR)(KR)[b]
	KRX_{10-12}K(KR)X(K/R)[b]

[a]Sequence representation is as follows: ($^\wedge$DE), any amino acid exept Asp or Glu; X_{10-12}, any 10-12 amino acids.

[b]For more optimal patterns, acidic residues should be rich in the central linker region and rare in the terminal linker region, whereas basic and hydrophobic residues should be rare in the central region and proline-rich in the terminal region.

Table 1. Consensus sequence of six classes of importin α-dependent NLS

2.3. Drug-protein interaction

Despite the introduction of newly developed drugs, such as lenalidomide and bortezomib, multiple myeloma is still difficult to treat and patients have a poor prognosis. In order to find novel drugs that are effective for multiple myeloma, we tested the antitumor activity of 29 phthalimide derivatives against several multiple myeloma cell lines. Among these derivatives, 2-(2,6-diisopropylphenyl)-5-amino-1H-isoindole-1,3-dione (TC11) was found to be a potent inhibitor of tumor cell proliferation and an inducer of apoptosis *via* activation of caspase-3, 8 and 9. This compound also showed *in vivo* activity against multiple myeloma cell line KMS34 tumor xenografts in ICR/SCID mice. To identify TC11-binding proteins, we used the IVV method [1-3,15]. We first prepared a cDNA library derived from KMS34 cells, because our data suggested that KMS34 cells were the most sensitive to TC11. As a bait, biotinylated TC11 was immobilized on a microfluidic chip and TC11-binding proteins were selected. Although the 4-amino group of TC11, which was experimentally inferred to be critical for the activity, was biotinylated *via* a linker, the biotinylation hardly affect the antitumor activity. Among 11 candidate TC11-binding proteins identified by the IVV method after 4 rounds of selection, we focused on the nucleolar phosphoprotein nucleophosmin (NPM). Sequencing revealed that three selected NPM clones, designated 1–183 NPM, encoded the 183 NH_2-terminal amino acids of NPM, which include the oligomerization domain and a part of the histone-binding domain. The enrichment efficiency of the NPM clones was confirmed to be 10^4-fold after 4 rounds of selection by RT-PCR. NPM is a multifunctional protein involved in both tumorigenesis and tumor suppression [41]; for example, it regulates cell proliferation and centrosome dupulication [42] and stabilizes oncoprotein Myc [43] and tumor-suppressor protein p53 [44]. Therefore, we hypothesized that NPM is involved in TC11-induced apoptosis of tumor cells. Immunofluorescence and NPM-knockdown studies in HeLa cells suggested that TC11 inhibits centrosomal clustering by inhibiting the centrosomal-regulatory function of NPM, thereby inducing multipolar mitotic cells, which undergo apoptosis. NPM may become a novel target for development of antitumor drugs active against multiple myeloma [28].

We screened 46 novel anilinoquinazoline derivatives for activity to inhibit proliferation of a panel of human cancer cell lines. Among them, Q15 showed potent *in vitro* growth-inhibitory activity towards cancer cell lines derived from colorectal cancer, lung cancer and multiple myeloma. It also showed antitumor activity towards multiple myeloma KMS34 tumor xenografts in lcr/scid mice *in vivo*. Unlike the known anilinoquinazoline derivative gefitinib, Q15 did not inhibit cytokine-mediated intracellular tyrosine phosphorylation. To elucidate the mechanism through which Q15 inhibits proliferation of tumor cells, we set out to identify Q15-binding proteins by means of the IVV method [1-3,15]. We prepared a cDNA library derived from total RNA of human colon carcinoma SW480 cells, because, like other tumor cells, SW480 cells were sensitive to Q15. Proteins that bind to biotinylated Q15 immobilized on beads were selected using the IVV method. From the library obtained after 5 rounds of selection, we analyzed the DNA sequences of 100 clones. Among them, we obtained six clones of a fragment of the Luzp5/NCAPG2 gene encoding $hCAP-G2_{262-476}$ containing the HEAT (Huntingtin, elongation factor 3, a subunit of protein phosphatase 2A, TOR lipid kinase) repeat domain. Although three other clones were obtained redundantly, they were confirmed to be false-positive clones by means of binding assay (data not shown). hCAP-G2 is a subunit of condensin II complex [45,46], which is regarded as a key player in mitotic chromosome condensation [47]. Immunofluorescence study indicated that Q15 compromises normal segregation of chromosomes, and therefore might induce apoptosis. Thus, our results indicate that hCAP-G2 is a novel therapeutic target for development of drugs active against currently intractable neoplasms [29].

3. Application of the IVV method for comprehensive interactome network analyses

Interactome networks are essential for complete systems-level descriptions of cells. Large-scale protein-protein interactions (PPIs) are integral in the analysis of topological and dynamic features of interactome networks. We introduce large-scale interactome network analyses using a combination of the IVV method and a biorobot for 50 human transcription factors (TFs).

Comprehensive analysis of PPIs is an important task in the field of proteomics, functional genomics and systems biology. PPIs are usually analyzed by means of biochemical methods such as pull-down assay and co-immunoprecipitation, yeast two-hybrid (Y2H) assay [35] and phage display [36]. Recently, the combined use of mass spectrometry (MS) with an affinity tag [48] has made biochemical methods more comprehensive and reliable. However, the testable interaction conditions are restricted by the properties of the biological sources. The Y2H assay is one of the major tools used in the discovery and characterization of PPIs [49]. However, the results of Y2H analyses often include many false positives due to auto-activating bait or prey fusion proteins [50] and interactions of proteins that are toxic to yeast cells cannot be examined. Phage display, the most widely used display technology [51], is an effective alternative, because the interactions between libraries and target proteins occur *in vitro*, allowing optimal conditions to be used for many different target proteins. However, the detectability of very low copy number proteins by phage display is still limited, because phage libraries are produced in living bacteria [52]. Totally *in vitro* display technologies such as ribosome display [53], the IVV method [1-3,15] and DNA display [54] can circumvent the above difficulties, because they do not need living cells. As a model bait protein, we chose the basic leucine zipper (bZIP) domain of Jun protein, an important transcription factor, to screen Jun interactors from a mouse brain cDNA library. By performing iterative affinity selection and sequence analyses, we selected 16 novel Jun-associated protein candidates in addition to four known interactors. By means of real-time PCR and pull-down assay, 10 of the 16 newly discovered candidates were confirmed to be direct interactors with Jun *in vitro*. Furthermore, interaction of 6 of the 10 proteins with Jun was observed in cultured cells by means of co-immunoprecipitation and observation of subcellular localization. These results demonstrate that this *in vitro* display technology is effective for the discovery of novel protein–protein interactions and can contribute to the comprehensive mapping of protein–protein interactions.

Furthermore, in a single experiment using bait Fos, more than 10 interactors, including not only direct, but also indirect interactions, were enriched. Further, previously unidentified proteins containing novel leucine zipper (L-ZIP) motifs with minimal binding sites identified by sequence alignment as functional elements were detected as a result of using a randomly primed cDNA library. Thus, we consider that this simple IVV selection system based on cell-free cotranslation could be applicable to high-throughput and comprehensive analysis of PPI and complexes in large-scale settings involving parallel bait proteins.

Interactome networks are essential for complete systems-level descriptions of cells. Large-scale PPIs are integral in the analysis of topological and dynamic features of interactome networks [55]. Several attempts to collect large-scale PPI data have been initiated using various model organisms [56-61] and were subsequently conducted in humans [62-64]. Traditionally, protein interaction data are collected using high-throughput *in vivo* expression tools based on the yeast two hybrid (Y2H; [49]) and tandem affinity purification-mass spectrometry (TAP-MS; [65]) methods. Experiments of this nature have provided large-scale PPI data, but they have only generated information on interacting partners, without considering binding domains in detail. In the field of systems biology, a further understanding of cellular networks will require more complete data sets describing the underlying physical interactions between cellular components. Thus, it is important to identify not only the binding partners, but also the interacting domains at the amino acid level [66]. In fact, the idea of mapping the interacting regions (IRs) involved in a PPI has been previously suggested in connection with several large-scale screens [67]. Our IVV method of analyzing PPIs [15,16] is well suited for large-scale, high-throughput mRNA display of the domain-based interactome using a randomly primed cDNA library, and we were able to achieve the first large-scale mapping of human IR data at the domain level for TF-related protein complexes. Functional domains were easily extracted based on the identified sequences using a randomly primed prey library as a non-biased representation [15]. Bait mRNA templates were prepared *in vitro* [15], and large-scale IVV method was performed using a biorobot that can simultaneously execute up to 96 selections. Fifty human TF-related proteins were used as bait, and a human brain cDNA library was used as prey. A modified high-throughput version of IVV selection was employed [15]. Integration of large-scale PPI data with other data sets, such as 3D structural information [68] and expression data [69], is necessary to identify the possible functions of interaction networks [68]. Large-scale IR data sets are expected to reflect functional domains and to indicate the biological roles of the network without the need to integrate additional data. We confirmed the reliability and accuracy of our data by performing pull-down assays [15] and by examining the overlap between our results and known PPI domains using Pfam search [70]. The core data set (966 IRs; 943 PPIs) displayed a verification rate of 70%. Analysis of the IR data set revealed the existence of IRs that interact with multiple partners (Fig. 4). Furthermore, these IRs were preferentially associated with intrinsic disorder. This finding supports the hypothesis that intrinsically disordered regions play a major role in the dynamics and diversity of TF networks through their ability to structurally adapt to and bind with multiple partners. Accordingly, this domain-based interaction resource represents an important step in refining protein interactions and networks at the domain level and in associating network analysis with biological structure and function.

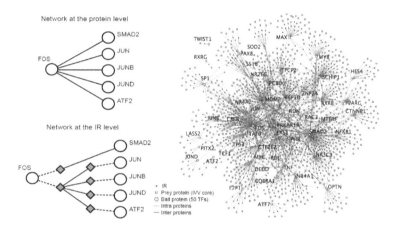

Figure 4. A transcription factor network at the interaction region(IR) level developed using IVV data. Graphic expression of the PPI network at the IR level(right). Interacting interfaces of the proteins, determined as IRs by IVV experiments, are drawn on the graph as diamond-shape nodes (IR nodes). Broken and solid lines indicate 'intra-' and 'inter-' protein edges, respectively. The graph contains 1,572 nodes (842 IR nodes and 730 protein nodes) and 842 intra-protein edges. Note that overlapping IRs are merged into a single node in the constructed network. An example of an underlying network graph at the IR level. Graphical expression of the FOS network at the protein level (upper left). PPIs are simply expressed by nodes indicating proteins and edges that connect them. Graphical expression of the FOS network at the IR level (lower left). A leucine zipper region of the FOS protein exclusively interacts with leucine zipper regions of other proteins (JUN, JUNB, JUND and ATF2). In addition, a region distinct from the leucine zipper in the FOS protein interacts with SMAD2.

4. Ultrahigh enrichment of antibodies by the IVV method on a microfluidic chip

Rapid preparation of monoclonal antibodies with high affinity and specificity is required in diverse fields from fundamental molecular and cellular biology to drug discovery and diagnosis [71]. In addition to classical hybridoma technology, *in vitro* antibody-display technologies [30,53,72-75] are powerful approaches for isolating single chain Fv (scFv) antibodies from recombinant antibody libraries. However, these display techniques require several rounds of affinity selection (typically, the library size is $10^7 \sim 10^{12}$, while the enrichment efficiency is $10 \sim 10^3$-fold per round). Recently, microfluidic systems have been developed for high-throughput protein analysis [76], since they offer the advantages of very low sample volumes, rapid analysis, and automated recovery of captured analytes for further characterization. However, there have been few attempts to combine microfluidic systems with *in vitro* antibody-display technologies so far. We showed that a microfluidic system can be combined with the IVV method [1-3,15] and employed for scFv selection from naïve and randomized scFv libraries with ultrahigh efficiency of 10^6- to 10^8-fold per round [31].

4.1. *In vitro* selection of antibodies from a naïve scFv library

The IVV selection of scFv was performed on a Biacore microfluidic chip. Since the diversity of the mouse scFv library prepared from mouse spleen poly A^+ RNAs is estimated to be $10^6 \sim 10^8$, while IVV method [15] allows screening of $\sim 10^{12}$ molecules, we also introduced random point mutations into the scFv library. We chose p53 (human tumor suppressor protein) and MDM2 (human murine double minute) proteins as model antigens that were immobilized on the Biacore sensor chip. The selection experiment was performed on the microfluidic chip, and selected scFv genes were amplified by reverse transcription (RT)-PCR and identified by cloning and sequencing. Unexpectedly, the recovered anti-p53 and anti-MDM2 scFv sequences converged on a single sequence and two sequences, respectively, after only two rounds of selection. These clones showed high affinity, but also low antigen-specificity, in pull-down assays, and so we examined the clones obtained after a single round of selection in each case. When the binding activities of 29 (anti-p53) and 20 (anti-MDM2) clones with distinct sequences were examined by means of pull-down assays, P1-93 and M1-19 showed high specificity against the respective antigens among p53, MDM2 and BSA (Fig. 5A). The amino-acid sequences of P1-93 and M1-19 are shown in Fig. 5B. In competitive ELISA, both clones dose-dependently inhibited the ELISA signal (Fig. 5C), and Scatchard plots revealed that the K_Ds of P1-93 and M1-19 were 22 nM and 5.9 nM, respectively. The K_Ds of P1-93 and M1-19 were also determined by surface plasmon resonance (SPR) as 12 nM and 4.3 nM, respectively (Fig. 5D). The values obtained by the two different methods are similar.

4.2. *In vitro* evolution of scFv

Further, we performed *in vitro* evolution of scFv with higher affinity against MDM2 from a randomly mutated M1-19 scFv library. We applied on-rate or off-rate selection as a selection pressure for *in vitro* affinity maturation with the Biacore instrument: the on-rate selection was performed by controlling flow rate, and the off-rate selection was carried out by using a prolonged washing process on the sensor chip. After one round of selection, the recovered scFv genes were cloned and sequenced, and the K_Ds were evaluated by competitive ELISA (Figs. 6A and 6D). We obtained four mutants with higher affinity for MDM2 ($K_D = 0.7$-3.8 nM) than the progenitor M1-19 from 22 distinct clones. The strongest binder, M1-19a, was con-

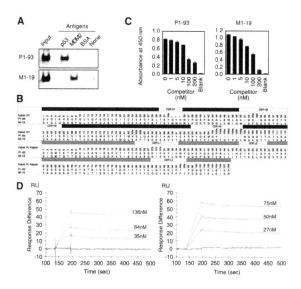

Figure 5. The selected scFvs anti-p53 P1-93 and anti-MDM2 M1-19. (A) Pull-down assays of the anti-p53 scFv P1-93 (top) and anti-MDM2 scFv M1-19 (bottom) using p53-, MDM2- or BSA-immobilized beads. Recovered scFv with FLAG-tag was detected with the anti-FLAG antibody. (B) Predicted amino-acid sequences of the V_H (black bar) and V_L (gray bar) regions of anti-p53 scFv P1-93 and anti-MDM2 scFv M1-19. (C) Competitive ELISA. P1-93 or M1-19 was pre-incubated with a competitor (0~200 nM free antigen) and allowed to bind to antigen-immobilized plates. After washing, remaining scFvs were detected with the anti-T7 tag antibody. (D) Biacore sensorgrams of the purified P1-93 (left: 31 kDa) and M1-19 (right: 32 kDa) using a p53- (blue lines: 55 kDa) or MDM2-immobilized (red lines: 66 kDa) sensor chip. The measurements were performed under conditions of 450 RU of the ligand and at a flow rate of 60 μl/min. To determine dissociation constants, three different concentrations (35, 64, and 136 nM for P1-93 and 27, 50, and 75 nM for M1-19) of the monomeric scFvs were injected.

firmed to have a higher on-rate and lower off-rate than M1-19 by SPR (Fig. 6B) and was also confirmed to recognize only the antigen protein MDM2 in crude cell lysates by Western blotting (Fig. 6C). These results indicated that the selected scFv had high enough affinity and specificity for practical use. Although the mutations of the selected scFvs were distributed among the whole sequences and no "consensus" mutations were identified, the mutation Y100bH within V_H CDR3 may contribute to the improved affinity and specificity, because this region is usually important for binding with antigens.

4.3. Ultrahigh efficiency of protein selection

Surprisingly, our results indicated that positive clone(s) were efficiently enriched through only one or two rounds of selection from a large library containing ~10^{12} molecules, implying ultrahigh efficiency of the method. To estimate the enrichment efficiency, we performed model experiments using a mixture of two kinds of scFv genes. The P1-93 (anti-p53) or M1-19 (anti-MDM2) gene was mixed with an anti-fluorescein scFv gene [30]('Flu' as a negative control) at a ratio of $1:10^2$, $1:10^4$, $1:10^6$ or $1:10^8$, and subjected to one round of the IVV selection on the sensor

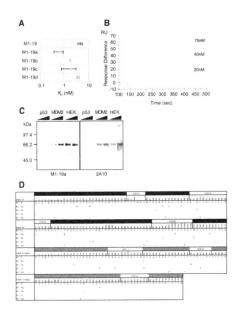

Figure 6. *In vitro* evolution of anti-MDM2 scFv M1-19. (A) Dissociation constants (K_Ds) of M1-19 (red circle) and the four mutant scFvs (green circles) were obtained by means of Scatchard plots of the competitive ELISA data. (B) Biacore sensorgrams of the mutant M1-19a (the amount of immobilized antigen was 450 RU). M1-19a had a higher on-rate and lower off-rate ($k_a = 2.5 \times 10^5$/Ms, $k_d = 8.6 \times 10^5$/s, $K_A = 3.0 \times 10^9$/M, $K_D = 0.34$ nM) than the progenitor M1-19 ($k_a = 5.5 \times 10^4$/Ms, $k_d = 2.3 \times 10^4$/s, $K_A = 2.4 \times 10^8$/M, $K_D = 4.3$ nM; see Fig. 5D: red lines on right side). (C) Recombinant p53 and MDM2 proteins (10 and 25 ng), and HEK-293 cell lysates (1 and 2.5 µg) were analyzed by Western blotting using M1-19a (left) or commercially available anti-MDM2 2A10 (right) as a control, respectively. (D) Predicted amino-acid sequences of the V_H (black bar) and V_L (gray bar) regions of the mutant scFvs M1-19a-d.

chip. The selection of the 1:10^6 mixture of P1-93:Flu genes and the 1:10^8 mixture of M1-19:Flu genes each resulted in a roughly 1:1 final gene ratio (Fig. 7A), indicating enrichment efficiencies of 10^6- and 10^8-fold per round, respectively. Furthermore, we confirmed that not only protein-protein (antigen-antibody) interactions, but also protein-DNA and protein-drug interactions were selected by our method with high enrichment efficiencies of >10^6-fold (Figs. 7B and 7C). Since the enrichment efficiencies of these model experiments with a usual agarose resin were only 10~10^3-fold per round, the enrichment efficiency was improved 10^3~10^5-fold over previous methods. Furthermore, we confirmed that the IVV system using a photocleavable linker between mRNA and protein is useful for *in vitro* selection of epitope peptides, recombinant antibodies, and drug-receptor interactions [77].

Although Biacore instruments have so far been utilized mainly to analyze biomolecular interactions by SPR, a few researchers have used this approach to fish for affinity targets from a randomized DNA library [78], phage-displayed protein libraries [79,80], or a ribosome-displayed antibody library [81]. However, the enrichment efficiency in these applications was not high. Why, then, was ultrahigh efficiency achieved in the present protein selection by IVV method? The IVV is a relatively small object pendant from its encoding RNA moiety, which is about ten times larger. Thus, nonspecific adsorption of RNA on solid surfaces is potentially significant. The matrix of the Biacore sensor chip consists of carboxymethylated dextran covalently attached to a gold surface and poorly binds nucleic acid molecules, since both materials are negatively charged. In contrast, phage display and ribosome display involve large protein moieties (coat proteins or ribosome), so the use of sensor chip may not improve the enrichment efficiency in these cases.

It should be noted that the ultrahigh enrichment efficiency made it difficult to set the number of selection rounds at a level that is appropriate to remove all non-binders as well as to pick all binders with various affinities from a library. If the number of selection rounds is too small, many negative sequences will be cloned; on the other hand, excess rounds of selection will yield only a single sequence with the highest affinity. In this study, we obtained 20-30 different sequences, including P1-93 and M1-19, with high antigen-specificity after a single round of selection, while we obtained only one or two negative sequences with high affinity but low antigen-specificity from the 10^6-10^8 library after two (probably excess) rounds of selection ($>10^{12}$-fold).

In summary, we achieved ultrahigh efficiencies (10^6~10^8-fold per round) of protein selection by IVV method with the microfluidic system. We obtained scFvs with high affinity and specificity from a naïve library by IVV selection for the first time. It took only three days to perform each selection experiment, including activity evaluation by ELISA. Although preparation of target materials of high quality is required, we anticipate this simple method to be a starting point for a versatile system to facilitate high-throughput preparation of monoclonal antibodies for analysis of proteome expression and detection of biomarkers, high-throughput

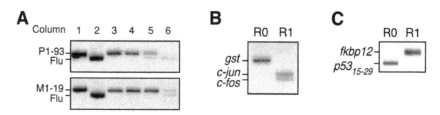

Figure 7. IVV selection of protein interactions on the Biacore sensor chip. (A) The P1-93 (912 bp: column 1 top) or M1-19 (936 bp: column 1 bottom) gene was mixed with an anti-fluorescein scFv gene [30](Flu: 888 bp: column 2) at a ratio of 1:10^2 (column 3), 1:10^4 (column 4), 1:10^6 (column 5) or 1:10^8 (column 6). The mixtures were subjected to IVV selection on the sensor chip conjugated with antigens p53 (top) or MDM2 (bottom). The RT-PCR products amplified from fractions after one round of selection were analyzed by agarose gel electrophoresis. (B) *In vitro* selection of protein-DNA interactions was performed using a mixture of three genes with N-terminal T7 tag and C-terminal FLAG-tag coding sequences: c-fos (349 bp), c-jun (394 bp), gst (597 bp). The template RNAs of c-fos, c-jun and gst (negative control) were mixed at a ratio of 1:1:10^6. The mixtures were translated and the resulting IVV libraries were selected on the sensor chip conjugated with bait DNA (AP-1)[20]. The RT-PCR products amplified from fractions before (R0) or after one round (R1) of selection were analyzed by agarose gel electrophoresis. (C) *In vitro* selection of protein-drug interactions was performed using a mixture of two genes with N-terminal T7 tag and C-terminal FLAG-tag coding sequences: fkbp12 (448 bp) and a p53 (15-29 aa) fragment (175 bp). The template RNAs of fkbp12 and the p53 fragment (negative control) were mixed at a ratio of 1:10^6. The mixtures were translated and the resulting IVV libraries were selected on the sensor chip conjugated with bait drug (FK506)[71]. The RT-PCR products amplified from fractions before (R0) or after one round (R1) of selection were analyzed by agarose gel electrophoresis.

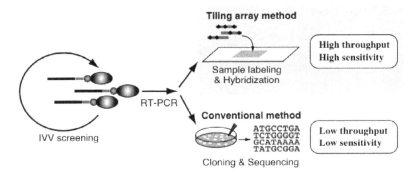

Figure 8. Schematic representation of the tiling array method and a conventional method for sequence determination of cDNAs from IVV screening.

analysis of protein-protein, protein-DNA and protein-drug interactions in proteomic and therapeutic fields, and rapid evolution of novel artificial proteins from large randomized libraries that often require ten or more rounds of selection.

5. Highly sensitive, high-throughput cDNA tiling arrays for detecting protein interactions selected by the IVV method

The most serious bottleneck in the IVV method has been in the final decoding step to identify the selected protein sequences. This step is usually achieved by cloning in bacteria and DNA sequencing using Sanger sequencers, but the following difficulties arise: 1) Only a limited number of clones can be analyzed, and thus positive candidates whose contents in the selected library are less than a threshold determined by the number of analyzed clones are lost as false negatives. 2) Positive sequences with low contents in a library can be enriched by iterative rounds of affinity selection, but lower-affinity binders compete with higher-affinity binders and therefore drop out of the screening. 3) DNA fragments which are injurious to cloning hosts, e.g., cytotoxic sequences, may be lost. 4) Cloning and sequencing of a huge number of copies of selected sequences is redundant, cost-ineffective, and time-consuming.

A DNA microarray is an efficient substitute for the cloning and sequencing processes to overcome the above limitations (Fig. 8). The combined use of a tiling array [82] representing ORF sequences with the IVV method would provide a completely *in vitro* platform for highly sensitive and parallel analysis of protein interactions. It should be possible to detect enrichment of cDNA fragments of selected candidates even with low contents or low affinity. However, for the analyses, the tiling arrays should be custom-designed specifically for the IVV screening, because the DNA fragments from the IVV method have unique characteristics, e.g., short length and functional region-concentrated distribution [16]. In this chapter, we introduce a highly sensitive, high-throughput protein interaction analysis procedure combining the IVV method with tiling arrays.

First, we designed a custom oligo DNA microarray as follows: 1) Oligonucleotide probes of 50-mer in length were used. This is the preferred length for microarray probes, because shorter probes result in low sensitivity and longer probes produce non-specific signals [83]. 2) There should be no gaps between the probes. A contiguous linear series of data is required to recognize a signal peak in the algorithm for tiling array analysis, as described below, so the probes must be densely arranged. 3) mRNA sequences were employed for the tiling array. Only coding regions are required for the purpose of protein-interaction analysis, so other genomic sequences, e.g., introns, control regions and non-coding RNAs, were not employed.

Second, we also improved the method for labeling of cDNA samples. Usually, double-stranded DNA samples for a tiling array analysis are labeled by using random primers [84]. However, cDNA fragments selected from a randomly fragmented cDNA library [16] seem to be too short for efficient labeling by random priming. Indeed, in a test analysis with a tiling array using the random priming labeling method, we failed to detect any of the previously detected positive controls. Therefore we employed another labeling procedure [85], in which sense-strand-labeled

RNAs were produced by one-step *in vitro* transcription using a SP6 promoter attached to cDNA fragments from IVV screening.

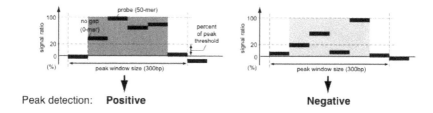

Peak detection: **Positive**   **Negative**

Figure 9. Windowed threshold detection algorithm for signal peak detection.

Third, we developed a detection algorithm for specific signal peaks from raw data. After iterative rounds of IVV screening, the resulting cDNA libraries in the presence and absence of bait protein, called bait (+) and bait (-) library, respectively, are labeled with fluorescence dyes using the above method, and hybridized separately. The ratios of the signal intensities from the experiments in the presence and absence of bait were calculated. Next, we searched for signal peaks in the data using the "windowed threshold detection" algorithm (Fig. 9). This algorithm looks for at least four data points that are above a threshold value within a window. These points were grouped together and presented as a peak. We used the following parameters in the algorithm: peak window size, 300 bp; percent of peak threshold, 20% of maximum data in each mRNA sequence. The value of each peak was the maximum value of the data points in that peak. Only reproducible peaks in the duplicated data were collected as candidates.

As an actual model study, we performed protein-protein interaction screening for mouse Jun protein [86], a transcription factor containing a bZIP domain, using the combined IVV and tilling array method. For this study, we constructed a novel custom microarray containing ~1,600 ORF sequences of known and predicted mouse transcription-regulatory factors (334,372 oligonucleotides) [16,87,88] to analyze cDNA fragments from IVV screening for Jun-interactors, and named it the Transcription-Factor Tiling (TFT) array. From the 5th-round DNA library of the IVV screening in the presence and absence of a bait Jun protein, we obtained labeled RNAs and hybridized them onto the TFT array [89].

Positive signal peaks were collected using the windowed threshold detection algorithm; the total number of peaks was 647 on 545 mRNA sequences (some of the mRNA sequences included multiple peaks) [89]. An example is shown in Fig. 10. To distinguish between true positives and false positives, specific enrichment of the selected candidate was validated by real-time PCR. Among the top 10 percent of the peaks (64 regions), specific enrichment of 35 peaks was confirmed in the screening (white bars in Fig. 11A). The data indicate that the appropriate threshold for distinguishing between true positives and noise in the microarray signal is a signal ratio of 3~4. The 35 candidates identified in the present study include all of the 20 Jun-interactors identified in our previous studies using conventional cloning and sequencing [16,88]. Furthermore, the 35 candidates include eight well-known Jun-associated proteins, which is double the number in the previous study, in which four known Jun-interactors were obtained (white bars of Fig. 11B) [16,88,89]. In other words, 15 proteins including four known Jun-interactors were newly detected using the TFT arrays.

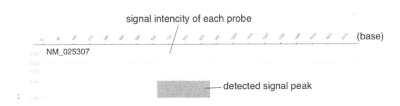

Figure 10. An example of probe signals and a detected signal peak on a gene.

Finally, we used *in vitro* pulldown assay and the surface plasmon resonance method to confirm the physical association of the 11 newly discovered candidates with Jun. As a result, ten of the 11 tested candidates exhibited specific interaction with the bZIP domain of Jun. Although most of the above tested interactions seem to be very weak, we considered that the interactions are true positives, because all of the candidates except for one contain leucine-heptad repeats in the selected regions, and such repeats are an important motif for heterodimerization with Jun [89].

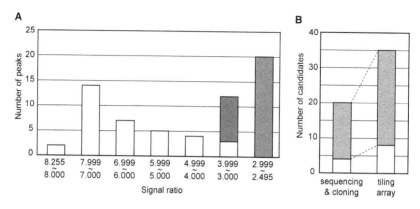

Figure 11. Data from the TFT array. (A) The top 10% of candidates were confirmed by real-time PCR. White and gray indicate numbers of enriched and non-enriched candidates, respectively. (B) Numbers of known (white) and newly selected (gray) proteins from conventional sequencing and the TFT arrays.

Previous studies and our survey revealed that the cDNA library used in this screening contained 29 known Jun-interactors [89,90]. Of these proteins, four (14%) and eight (28%) were detected by conventional sequencing and by the TFT array method, respectively. Thus, the TFT array method provides a remarkable increase in the number of identified interactors and this confirms the value of our new methodology as a screening tool for protein interactions. While the coverage was increased considerably, the accuracy did not decrease. Specifically, the number of false positives did not increase: the rates of confirmation of proteins by *in vitro* pull-down assays in the previous and present studies were 75% and 74%, respectively [90]. Undetected remaining interactors were considered to be false negatives. Mismatching of the selection conditions, e.g., salts, detergents, and pH, or the bait construct, e.g., length, region, and tags, might inhibit these interactions.

For quantitative analysis, the abundance ratios of 35 specifically selected candidates in the initial and screened cDNA libraries were determined by real-time PCR, and the enrichment rates (abundance ratio in the 5th round library per that in the initial library) were also calculated [88,89]. The abundance of the 15 newly found candidates (excluding four cases) was less than the theoretical threshold determined from the results of our previous study (an analysis of 451 clones). In order to detect the least abundant candidate (1.3×10^{-4}% of the screened cDNA library) by cloning and sequencing, it would have been necessary to analyze at least 1.0×10^6 clones. These results indicate that our new method is more sensitive, higher-throughput and more cost-effective than the previous method.

From the standpoint of the detection sensitivity, the combined use of the IVV method with tiling arrays provides an extremely sensitive method for protein-interaction analysis, because even a very weakly expressed target could be detected in this study. In the cDNA library before IVV screening, the content of fragments of the selected region of the least abundant known Jun-binder was 1.2×10^{-7}%. If one mRNA molecule existed per cell, the content of a fragment of the gene would be about 1.2×10^{-5} to 5.9×10^{-5}% (we employed reported parameters for this calculation [91]). Thus, the content of the least abundant mRNA in the initial library corresponds to about one molecule per 20 to 100 cells. This suggests that this gene is expressed at a very low level in a cell type that is a minor component of the tested tissue. It is noteworthy that targets expressed at such low levels can be detected without the need for a cell purification procedure, e.g., collection of somatic stem cells by flow cytometry. The high sensitivity of our method may allow access to targets which would be hard to analyze with other existing tools, such as the TAP method [48].

In summary, we have applied tiling array technology, which has previously been used for ChIP-chip assays and transcriptome analyses, to protein-interaction analysis with the IVV method. Compared with previous results obtained with cloning and sequencing, the use of the tiling array greatly increased sensitivity. This method can detect targets expressed at extremely low levels. This highly sensitive and reliable method has the potential to be used widely, because the tiling array approach can easily be extended to a genome-wide scale, even though the search space is limited in tiled sequences.

6. Conclusion

We have developed an mRNA display technology, named the *in vitro* virus (IVV) method, as a stable and efficient tool for analyzing various protein functions. The IVV method is applicable for exploring protein complexes, transcription factors, RNA-binding proteins, bioactive peptides, drug-target proteins and antibodies, as well as *in vitro* protein evolution from random-sequence and block-shuffling libraries. We further developed a large-scale and high-throughput IVV screening system utilizing a biorobot, microfluidic tip, and tiling array. Here we reviewed applications of the IVV method for protein functional analyses.

Acknowledgements

We thank Nobuhide Doi, Etsuko Miyamoto-Sato, Hideaki Takashima, Shunichi Kosugi, Masamichi Ishizaka, Toru Tsuji, Nobutaka Matsumura, Seiji Tateyama, Shigeo Fujimori, Hironobu Kimura, Isao Fukuda, Hirokazu Shiheido, Ichigo Hayakawa, Hiroaki Genma, Mayuko Tokunaga, Yuko Yonemura-Sakuma, Yoko Ogawa, Kazuyo Masuoka, Naoya Hirai and Masako Hasebe who are collaborators in the studies introduced in this review. We also thank Profs. Hideyuki Saya, Yutaka Hattori, Masaru Tomita and Taketo Yamada, and Drs Tatsuya Hirano and Takao Ono for valuable advice and discussion.

Author details

Noriko Tabata, Kenichi Horisawa and Hiroshi Yanagawa

Department of Biosciences and Informatics, Faculty of Science and Technology, Keio University, Yokohama, Japan

References

[1] Nemoto N, Miyamoto-Sato E, Husimi Y, Yanagawa H. *In vitro* virus: bonding of mRNA bearing puromycin at the 3'-terminal end to the C-terminal end of its encoded protein on the ribosome *in vitro*. FEBS Lett 1997;414: 405-408.

[2] Miyamoto-Sato E, Nemoto N, Kobayashi K, Yanagawa H. Specific bonding of puromycin to full-length protein at the C-terminus. Nucleic Acids Res 2000;28: 1176-1182.

[3] Miyamoto-Sato E, Takashima H, Fuse S, Sue K, Ishizaka M, et al. Highly stable and efficient mRNA templates for mRNA-protein fusions and C-terminally labeled proteins. Nucleic Acids Res 2003;31: e78.

[4] Nemoto N, Miyamoto-Sato E, Yanagawa H. Fluorescence labeling of the C-terminus of proteins with a puromycin analogue in cell-free translation systems. FEBS Lett 1999; 462: 43-46.

[5] Doi N, Takashima H, Kinjo M, Sakata K, Kawahashi Y et al. Novel fluorescence labeling and high-throughput assay technologies for *in vitro* analysis of protein interactions. Genome Res 2002;12: 487-492.

[6] Yarmolinsky MB, De La Hara,G.L. Inhibition by puromycin of amino acid incorporation into protein. Proc Natl Acad Sci USA 1959;45:1721-1729.

[7] Takeda Y, Hayashi S, Nakagawa H, Suzuki F. The effect of puromycin on ribonucleic acid and protein synthesis. J Biochem 1960;48:169-177.

[8] Nemeth AM, de la Haba GL. The effect of puromycin on the developmental and adaptive formation of tryptophan pyrrolase. J Biol Chem 1962;237:1190-1193.

[9] Nathans D. Puromycin inhibition of protein synthesis: Incorporation of puromycin into peptide chains. Proc Natl Acad Sci USA 1964;51:585-592.

[10] Traut RR, Monro RE. The puromycin reaction and itsw reaction to protein synthesis. J Mol Biol 1964;10:63-72.

[11] Steiner G., Kuechler E, Barta A. Photo-affinity labelling at the peptidyl transferase centre reveals two different positions for the A- and P-sites in domain V of 23S rRNA. EMBO J 1988;7:3949-3955.

[12] Kudlicki W, Odom OW, Kramer G, Hardesty B. Activation and release of enzymatically inactive, full-length rhodanese that is bound to ribosomes as peptidyl-tRNA. J Biol Chem 1994;269:16549-16553.

[13] Wolin SL, Walter P. Ribosome pausing and stacking during translation of a eukaryotic mRNA. EMBO J 1988;7:3559-3569.

[14] Bjornsson A, Isaksson LA. Accumulation of a mRNA decay intermediate by ribosomal pausing at a stop codon. Nucleic Acids Res 1996;24:1753-1757.

[15] Miyamoto-Sato E, Ishizaka M, Horisawa K, Tateyama S, Takashima H, et al. Cell-free cotranslation and selection using *in vitro* virus for high-throughput analysis of protein-protein interactions and complexes. Genome Res 2005;15:710-717.

[16] Horisawa K, Tateyama S, Ishizaka M, Matsumura N, Takashima H et al.: *In vitro* selection of Jun-associated proteins using mRNA display. Nucleic Acids Res 2004;32: e169.

[17] Kawahashi Y, Doi N, Takashima H, Tsuda C, Oishi Y et al. *In vitro* protein microarrays for detecting protein-protein interactions: Application of a new method for fluorescence labeling of proteins. Proteomics 2003;3:1236-1243.

[18] Miyamoto-Sato E, Ishizaka M, Fujimori S, Hirai N, Masuoka K et al. A comprehensive resource of interacting protein regions for refining human transcription factor networks: Domain-based interactome. PLoS ONE 2010;5: e9289.

[19] Suzuki H, Alistair RRF, Nimwegen E, Daub CO, Balwierz PJ. et al. The transcriptional network that controls growth arrest and differentiation in a human myeloid leukemia cell line. Nat Genet 2009;41:553-562.

[20] Tateyama, S, Horisawa, K, Takashima, H, Miyamoto-Sato, E, Doi, N et al. Affinity selection of DNA-binding protein complexes using mRNA display. Nucleic Acids Res 2006;34: e27.

[21] Horisawa K, Imai T, Okano H, Yanagawa H. 3′-Untranslated region of doublecortin mRNA is a binding target of the Musashi1 RNA-binding protein. FEBS Lett 2009;583:2429-2434.

[22] Horisawa K, Imai T, Okano H, Yanagawa H. The Musashi family RNA-binding proteins in stem cells. Biomol Concepts 2010;1:59-66.

[23] Shiheido H, Takashima H, Doi N, Yanagawa H. mRNA display selection of an optimized MDM2-binding peptide that potently inhibits MDM2-p53 interaction. PLoS ONE 2011;6:e17898.

[24] Matsumura N, Tsuji T, Sumida T, Kokubo M, Onimaru M et al. mRNA display selection of a high-affinity Bcl-X$_L$-specific binding peptide. FASEB J 2010;24:2201-2210.

[25] Kosugi S, Hasebe M, Matsumura N, Takashima H, Miyamoto-Sato E et al. Six classes of nuclear localization signals specific to different binding grooves of importinα. J Biol Chem 2009;284:478-485.

[26] Kosugi S, Hasebe M, Tomita M,Yanagawa H. Systematic identification of cell cycle-dependent nucleocytoplasmic shuttling proteins by prediction of composite motifs. Pro Natl Acad Sci USA 2009;106: 10171-10176.

[27] Kosugi S, Hasebe M, Entani T, Takayama S, Tomita M et al. Design of peptide inhibitors for importin α/β nuclear import pathway by activity-based profiling. Chem & Biol 2008;15:940-949.

[28] Shiheido H, Terada F, Tabata N, Hayakawa I, Matsumura N, et al. A phthalimide derivative that inhibits centrosomal clustering is effective on multiple myeloma. PLoS ONE 2012;7,e38878.

[29] Shiheido H, Naito Y, Kimura H, Genma H, Takashima H. An anilinoquinazoline derivative inhibits tumor growth through interaction with hCAP-G2, a subunit of condensing II. PLOS ONE 2012;7:e44889.

[30] Fukuda I, Kojoh K, Tabata N, Doi N, Takashima H et al. *In vitro* evolution of single-chain antibodies using mRNA display. Nucleic Acids Res 2006;34: e127.

[31] Tabata N, Sakuma Y, Honda Y, Doi N, Takashima H. et al. Rapid antibody selection by mRNA display on a microfluidic chip. Nucleic Acids Res 2009;37: e64.

[32] Pollack JR, Iyer VR. Characterizing the physical genome. Nature Genet 2002;32:515–521.

[33] Yu H, Luscombe NM, Quian J, Gerstein M. Genomic analysis of gene expression relationships in transcriptional regulatory networks. Trends Genet 2003;19: 422–427.

[34] Barski A, Frenkel B. ChIP display: novel method for identification of genomic targets of transcription factors.Nucleic Acids Res 2004;32, e104.

[35] Luo Y, Vijaychander S, Stile J, Zhu,L. Cloning and analysis of DNA-binding proteins by yeast one-hybrid and one-two-hybrid systems. Biotechniques 1996;20:564–568.

[36] Hagiwara H, Kunihiro S, Nakajima K, Sano M, Masaki H. et al. Affinity selection of DNA-binding proteins from yeast genomic DNA libraries by improved phage display vector. J Biochem 2002;132:975–982.

[37] Borghouts C, Kunz C, Groner B. Current strategies for the development of peptide-based anti-cancer therapeutics. J Peptide Sci 2005;11: 713–726.

[38] Bottger V, Bottger A, Howard SF, Picksley SM, Chene P. et al. Identification of novel mdm2 binding peptides by phage display. Oncogene 1996;13: 2141–2147.

[39] Vassilev LT, Vu BT, Graves B, Carvajal D, Podlaski F. et al. *In vivo* activation of the p53 pathway by small-molecule antagonists of MDM2. Science 2004;303: 844–848.

[40] Hu B, Gilkes DM, Chen J. Efficient p53 activation and apoptosis by simultaneous disruption of binding to MDM2 and MDMX. Cancer Res 2007;67:8810–8817.

[41] Grisendi S, Mecucci C, Falini B, Pandolfi PP. Nucleophosmin and cancer. Nat Rev Cancer 2006;6: 493–505.

[42] Okuda M. The role of nucleophosmin in centrosome duplication. Oncogene 2002;21: 6170–6174.

[43] Li Z, Boone D, Hann SR. Nucleophosmin interacts directly with c-Myc and controls c-Myc-induced hyperproliferation and transformation. Proc Natl Acad Sci USA 2008;105:18794–18799.

[44] Colombo E, Marine JC, Danovi D, Falini B, Giuseppe P. Nucleophosmin regulates the stability and transcriptional activity of p53. Nat Cell Biol 2002;4: 529–533.

[45] Ono T, Losada A, Hirano M, Myers MP, Neuwald AF, et al. Differential contributions of condensin I and condensin II to mitotic chromosome architecture in vertebrate cells. Cell 2003;115: 109–121.

[46] Hirano T. Condensins: Organizing and segregating the genome. Curr Biol 2005;15:R265–275.

[47] Hudson DF, Marshall KM, Earnshaw WC. Condensin: Architect of mitotic chromosomes. Chromosome Res 2009;17:131–144.

[48] Puig O, Caspary F, Rigaut G, Rutz B, Bouveret E. et al. The tandem affinity purification (TAP) method: a general procedure of protein complex purification. Methods 2001;24:218-229.

[49] Fields S, Song O. A novel genetic system to detect protein-protein interactions. Nature 1989;340:245-246.

[50] Vidalain PO, Boxem M, Ge H, Li S, Vidal, M. Increasing specificity in high-throughput yeast two-hybrid experiments. Methods 2004;32:363-370.

[51] Smith GP. Filamentous fusion phage: novel expression vectors that display cloned antigens on the virion surface. Science 1985;228:1315-1317.

[52] Bradbury AR, Marks JD. Antibodies from phage antibody libraries. J Immunol Methods 2004;290: 29-49.

[53] Hanes J, Pluckthun A. In vitro selection and evolution of functional proteins by using ribosome display. Proc Natl Acad Sci USA 1997; 94:4937-4942.

[54] Yonezawa M, Doi N, Kawahashi Y, Higashinakagawa T, Yanagawa, H. DNA display for in vitro selection of diverse peptide libraries. Nucleic Acids Res 2003;31: e118.

[55] Jeong H, Mason SP, Barabasi AL, Oltvai ZN. Lethality and centrality in protein networks. Nature 2001;411: 41-42.

[56] Uetz P, Giot L, Cagney G, Mansfield TA, Judson RS. et al. A comprehensive analysis of protein-protein interactions in Saccharomyces cerevisiae. Nature 2000;403:623-627.

[57] Ito T, Chiba T, Ozawa R, Yoshida M, Hattori M. et al. A comprehensive two-hybrid analysis to explore the yeast protein interactome. Proc Natl Acad Sci USA 2001;98:4569-4574.

[58] Gavin AC, Bosche M, Krause R, Grandi P, Marzioch M. et al. Functional organization of the yeast proteome by systematic analysis of protein complexes. Nature 2002;415:141-147.

[59] Giot L, Bader JS, Brouwer C, Chaudhuri A, Kuang B. et al. A protein interaction map of Drosophila melanogaster. Science 2003;302:1727-1736.

[60] Li S, Armstrong CM, Bertin N, Ge H, Milstein S. et al. A map of the interactome net-
      work of the metazoan *C. elegans*. Science 2004;303:540-543.

[61] Butland G, Peregrin-Alvarez JM, Li J, Yang W, Yang X, et al. Interaction network con-
      taining conserved and essential protein complexes in *Escherichia coli*. Nature
      2005;433:531-537.

[62] Rual JF, Venkatesan K, Hao T, Hirozane-Kishikawa T, Dricot A. et al. Towards a pro-
      teome-scale map of the human protein-protein interaction network. Nature
      2005;437:1173-1178.

[63] Stelzl U, Worm U, Lalowski M, Haenig C, Brembeck FH. et al. A human protein-pro-
      tein interaction network: a resource for annotating the proteome. Cell
      2005;122:957-968.

[64] Ewing RM, Chu P, Elisma F, Li H, Taylor P, et al. Large-scale mapping of human
      protein-protein interactions by mass spectrometry. Mol Syst Biol 2007;3: 89.

[65] Rigaut G, Shevchenko A, Rutz B, Wilm M, Mann M. et al. A generic protein purifica-
      tion method for protein complex characterization and proteome exploration. Nat Bio-
      technol 1999;17:1030-1032.

[66] Hakes L, Pinney JW, Robertson DL, Lovell SC. Protein-protein interaction networks
      and biology--what's the connection? Nat Biotechnol 2008;26:69-72.

[67] Fromont-Racine M, Rain JC, Legrain P. Toward a functional analysis of the yeast ge-
      nome through exhaustive two-hybrid screens. Nat Genet 1997;16:277-282.

[68] Kim PM, Lu LJ, Xia Y, Gerstein MB. Relating three-dimensional structures to protein
      networks provides evolutionary insights. Science 2006;314: 1938-1941.

[69] Han JD, Bertin N, Hao T, Goldberg DS, Berriz GF. et al. Evidence for dynamically or-
      ganized modularity in the yeast protein-protein interaction network. Nature
      2004;430:88-93.

[70] Finn RD, Mistry J, Schuster-Bockler B, Griffiths-Jones S, Hollich V. et al. Pfam: clans,
      web tools and services. Nucleic Acids Res 2006;34:D247-251.

[71] Hoogenboom, H.R. Selecting and screening recombinant antibody libraries. Nat Bio-
      technol 2005;23:1105-1116.

[72] Marks JD, Hoogenboom HR, Bonnert TP, McCafferty J, Griffiths AD. By-passing im-
      munization. Human antibodies from V-gene libraries displayed on phage. J. Mol. Bi-
      ol 1991;222:581-597.

[73] He M, Taussig, M.J. Antibody-ribosome-mRNA (ARM) complexes as efficient selec-
      tion particles for *in vitro* display and evolution of antibody combining sites. Nucleic
      Acids Res 1997;25:5132-5134.

[74] Reiersen H, Lobersli I, Loset GA, Hvattum E, Simonsen B. et al. Covalent antibody display--an *in vitro* antibody-DNA library selection system. Nucleic Acids Res 2005;33:e10.

[75] Boder ET, Midelfort KS,Wittrup KD. Directed evolution of antibody fragments with monovalent femtomolar antigen-binding affinity. Proc Natl Acad Sci USA 2000;97:10701-10705.

[76] Lion N, Rohner TC, Dayon L, Arnaud IL, Damoc E. et al. Microfluidic systems in proteomics. Electrophoresis 2003;24:3533-3562.

[77] Doi N, Takashima H, Wada A, Oishi Y, Nagano T. et al. Photocleavable linkage between genotype and phenotype for rapid and efficient recovery of nucleic acids encoding affinity-selected proteins. J Biotechnol 2007;131:231-239.

[78] Hao D, Ohme-Takagi M, Yamasaki K. A modified sensor chip for surface plasmon resonance enables a rapid determination of sequence specificity of DNA-binding proteins. FEBS Lett 2003;536:151-156.

[79] Yamamoto Y, Tsutsumi Y, Yoshioka Y, Nishibata T, Kobayashi K. et al. (2003) Site-specific PEGylation of a lysine-deficient TNF-alpha with full bioactivity. Nat Biotechnol 2003;21:546-552.

[80] Malmborg AC, Dueñas M, Ohlin M, Söderlind E, Borrebaeck CA. Selection of binders from phage displayed antibody libraries using the BIAcore™ biosensor. J. Immunol. Methods 1996;198:51-57.

[81] Yuan Q, Wang Z, Nian S, Yin Y, Chen G. et al. Screening of high-affinity scFvs from a ribosome displayed library using BIAcore biosensor. Appl. Biochem. Biotechnol 2009;152:224-234.

[82] Kapranov P, Cawley SE, Drenkow J, Bekiranov S, Strausberg RL, et al. Large-scale transcriptional activity in chromosomes 21 and 22. Science 2002;296(5569): 916-919.

[83] Nuwaysir EF, Huang W, Albert TJ, Singh J, Nuwaysir K. et al. Gene expression analysis using oligonucleotide arrays produced by maskless photolithography. Genome Res 2002 ;12:1749-1755.

[84] Kim TH, Barrera LO, Zheng M, Qu C, Singer MA. et al. A high-resolution map of active promoters in the human genome. Nature 2005;436:876-880.

[85] 't Hoen PA, de Kort F, van Ommen GJ, den Dunnen JT. Fluorescent labelling of cRNA for microarray applications. Nucleic Acids Res 2003;31: e20.

[86] Bohmann D, Bos TJ, Admon A, Nishimura T, Vogt PK. et al. Human proto-oncogene c-jun encodes a DNA binding protein with structural and functional properties of transcription factor AP-1. Science 1997;238:1386-1392.

[87] Gunji W, Kai T, Sameshima E, Iizuka N, Katagi H. et al. Global analysis of the expression patterns of transcriptional regulatory factors in formation of embryoid bodies

using sensitive oligonucleotide microarray systems. Biochem Biophys Res Commun 2004;325:265-275.

[88] Horisawa K, Doi N, Takashima H, Yanagawa H. Application of quantitative real-time PCR for monitoring the process of enrichment of clones on *in vitro* protein selection. J Biochem 2005;137:121-124.

[89] Horisawa K, Doi N, Yanagawa H. Use of cDNA tiling arrays for identifying protein interactions selected by *in vitro* display technologies. PLoS ONE. 2008;3:e1646.

[90] Chinenov Y, Kerppola TK Close encounters of many kinds: Fos-Jun interactions that mediate transcription regulatory specificity. Oncogene 2001;20:2438-2452.

[91] Campbell NA. Biology, 4th Ed. New York: The Benjamin/Cummings Publishing; 1996

Experimental Molecular Archeology: Reconstruction of Ancestral Mutants and Evolutionary History of Proteins as a New Approach in Protein Engineering

Tomohisa Ogawa and Tsuyoshi Shirai

Additional information is available at the end of the chapter

1. Introduction

The diversity of life on Earth is the result of perpetual evolutionary processes beginning at life's origins; evolution is the fundamental development strategy of life. Today, studies of gene and protein sequences, including various genome-sequencing projects, provide insight into these evolutionary processes and events. However, the sequence data obtained is restricted to extant genes and proteins, with the exception of the rare fossil genome samples [1, 2], for example Neanderthal [3], archaic hominin in Siberia [4, 5], and ancient elephants such as mastodon and mammoth [6]. The fossil record, and genome sequences derived from it, has the potential to elucidate ancient, extinct forms of life, acting as missing links to fill evolutionary gaps; however, the sequenced fossil genome is very limited, mainly due to the condition of samples and the challenges of preparing them. Discovering the forms of ancient organisms is one of the major purposes of paleontology, and is valuable in understanding of current life forms as these will be a reflection of their evolutionary history. However, the reconstruction of a living organism from fossils, which would be the ultimate paleontological methodology, is far beyond the currently available technologies, although there has recently been a report of the production of an artificial bacterial cell, using a chemically synthesized genome [7].

Meanwhile, for genes or the proteins they encode, it is already feasible to reconstruct their ancestral forms using phylogenetic trees constructed from sequence data; these techniques may well provide clues to the evolutionary history of certain extant genes and proteins with respect to their ancestors. Although phylogenetic analyses alone, or in combination with protein structure simulations, are useful to analyze structure-function relationships and evolutionary history [8], resurrected ancient recombinant proteins have the potential to provide more direct observations. Production of ancestral or ancient proteins can be achieved comparably easily due to developments in molecular biology and protein engineering techniques, which allow nucleotide or amino acid sequences to be synthesized. Ancestral proteins can be tested in the laboratory using biochemical or biophysical methods, for their activity, stability, specificity, and even three-dimensional structure. Thus, ancestral sequence reconstruction (ASR) has proved a useful experimental tool for studying the diverse structure and function of proteins [9]. To date, such 'experimental molecular archeology' using ASR has been applied to several enzymes [10-24], including photo-reactive proteins [25-37], nuclear receptor and transmembrane proteins [38-48], lectins [49-52], viral proteins [53, 54], elongation factor [55-57], paralbmin [58], in addition to a number of peptides [59,60] (Table 1).

In early studies, ASR experiments using the technique of molecular phylogeny were based on basic site-directed mutagenesis and used to investigate the functional evolution of proteins, including the convergent evolution of lysozyme in ruminant stomach environments and the adaptation of enzymes to alkaline conditions [10-12]. However, if ancestral sequences have been determined, the most straightforward method is to reconstruct the full-length ancestral protein in the laboratory. No fundamental differences exist between ancestor reconstruction and standard site-directed mutagenesis, other than the number of amino acids residues requiring mutation, which, in the case of ancestor reconstruction, might be spread over the entire sequence. At present, ASR can be achieved using commercially available *de novo* synthetic genes. Thus, 'experimental molecular archeology' by ancestral protein reconstruction using a combination of the technical developments in biochemistry, molecular biology, and bioinformatics can be exploited in both molecular evolutionary biology and protein engineering. In this chapter, we will provide an overview the experimental molecular archeology technique of ASR, and the case of ancestral fish galectins will be discussed in detail, based on our recent studies.

2. The early studies: Reconstruction of partial ancestors by site-directed mutagenesis

The first studies exploiting the idea of ancestral protein reconstruction used site-direct mutagenesis, in which a small number of amino acids were substituted to produce the anticipated ancestral status. These studies include the reconstruction of a ribonuclease (RNase) of an extinct bovid ruminant [10, 11], and the lysozymes from a game-bird using ancestral lysozyme reconstructions predicted by the MP (Maximum Parsimony) method [12]. Benner and colleagues reconstructed RNase of an extinct bovid ruminant [10], by predicting four sequences of ancestral RNases from five closely related bovids including ox, swamp buffalo, river buffalo, nilgai, and the primitive artiodactyl using the MP method [61, 62]. The ancestor closest to the extant ox protein was selected from the four probable ancestors as the target of the experiment as it contained a mutation of amino acid residue 35, located close to Lys41, which is known to be important for catalysis. Three ancestral mutants of the ox RNase (A19S, L35M, and A19S/L35M) were examined for their kinetic properties and the thermal stabilities against tryptic digestion. However, no significant difference was found between the ancestral

Ancestral proteins (family)	Methods/programs	Partial/ full length	Remarkable future	References
[Enzymes]				
RNases		point mutants		[10, 11]
Lysozyme				[12]
Alchole dehydrogenase	ML (PAML)	full length	similar to Adh1 than Adh2 accumulation of ethanol	[13]
Isopropylmalate dehydrogenase (IPMDH)		multiple-point mutants		[14-16]
Isocitrate dehydrogenase (ICDH)				[17]
				[18]
3-isopropylmalate dehydrogenase (LeuB)	ML/Bayesian	full length	ancestral 3D structures	[19]
Chymase	MP	full length	ancestor of α/β-chymase Ang II-forming activity	[20]
DNA gyrase	ML (Tree-Puzzle)	Partial: ATPase D	thermal stability	[21]
Glycyl-tRNA synthetase (GlyRS)	ML	multiple-point mut	Commonote, thermal stability	[22]
Sulfotransferases /Paraoxonase	ML (FastML)	multiple-point mut	Directed evolution ancestral 3D structure	[23]
Thioredoxin (Trx)			Precambrian enzyme (anoxygenic/oxygenic environment)	[24]
[Visual pigment proteins & Fluorescent proteins]				
Opsins (rhodopsin)	ML (nucleotide/amino acids/codon)		Archosaur ancestor	[25]
		multiple-point mut	UV pigment (SWS1)	[26]
	ML-based Bayesian	multiple-point mut	Rhodopsin (RH1)	[27]
		multiple- point mut	Red/green color vison	[28]
			Zebrafish RH2-1~4	[29]
	ML-based Bayesian (PAML), MP	multiple-point mut	Dim-light (deep-sea) vision	[30-32]
GFP (coral pigment)	ML (MrBayes 3.0)	full length	Fluorescence spectra evolved from green common ancestor convergent evolution/positive selection	[33-37]
[Receptors & transmembrane proteins]				
Nuclear receptors (NR) for steroid estrogen/androgen/progesterone/ glucocorticoid/mineralocorticoid receptors	MP/ML		ancestral 3D structures NR superfamily	[38-47]
Vacuolar H1-ATPase				[48]
[Carbohydrate binding proteins/Lectin]				
Galectins	ML/MP	full length	carbohydrate binding ancestral 3D structures	[49-51]
Tachlectin-2 (β-propeller lectin)		fragments	oligomeric assembly	[52]
[Viral proteins]				
Coxsackievirus B5 capsid	ML (PAML)	P1 region	Infectious activity, cell binding Cell tropism, antigenicity	[53]
Core protein PtERV1 p12-Capsid	MP	point mut	TRIM5α antiviral protein	[54]
[Others]				
Elongation factors (EF) Tu	ML (MOLPHY/PAML/ JTT/Dayhoff/WAG)	full length	thermostability phenotype genotype Hyperthermophiles	[55-57]
Parvalbumin (PVs)	ML (FastML)	point mutants	thermal adaptation	[58]
Allatostatins (ASTs)	ML (PAML)/ consider gap	peptide	juvenile hormone release inhibition	[59]
Glucagon-like peptide-1 (GLP-1)	ML (FastML) with JTTmatrix	peptide	receptor affinity, stability	[60]

ML: Maximum likelihood/bayesian, MP: Maximum Parsimony

Table 1. The experimental molecular archeology analysis using ancestral proteins

mutants and the modern ox RNase. The results suggested that these amino acid substitutions were evolutionarily neutral, although this conclusion is limited to the extent of the examined properties [11].

Malcolm et al. succeeded in identifying a non-neutral evolutionary pathway of game-bird lysozymes using ancestral lysozyme reconstructions predicted by the MP method [12]. Seven mutations in game-bird lysozyme proteins included combinations of residues Thr40, Ile55, and Ser91, which were anticipated to be Ser40, Val55 and Thr91, respectively, in ancestral molecules. The mutants were synthesized as possible intermediates in the evolutionary pathway of bird lysozyme and comparative molecular properties and crystal structures of these revealed that the thermostabilities of the proteins were correlated with the bulkiness of their side chains. The T40S mutant increased its thermostability by more than 3°C, allowing the conclusion that this mutation was non-neutral effect of natural selection.

Yamagishi and colleagues used ancestral protein reconstruction [14-16] to obtain direct evidence for the hypothesis that the common ancestor of all organisms was hyper-thermophilic [63]. Because the catalytic activities of 3-isopropylmalate dehydrogenase (IPMDH) and isocitrate dehydrogenase (ICDH) are similar to one another and their three-dimensional structures conserved, these proteins are diverged from an ancient common ancestor [64], of which sequence was inferred from a phylogenetic tree constructed from IPMDH and ICDH sequences from various species, including the thermophile (*Thermus thermophilus*) and the extreme thermophile (*Sulfolobus* sp. strain 7). Five of the seven ancestral mutants, in which substituted amino acids were located close to the substrate and cofactor-binding sites, demonstrated higher thermostability than wild type IPMDH from *Sulfolobus* sp. strain 7. These findings were taken to support the hypothesis of a hyperthermophile common ancestor. Moreover, the successful thermostabilization of ICDH [17] and Glycyl-tRNA synthetase [22] by ASR has been reported. Thus, the incorporation of ancestral residues into a modern protein can be used not only to test evolutionary hypotheses, but also as a powerful protein engineering technique for protein thermostabilization.

Recently, Whittington and Moerland reported that ASR analysis of parvalbumins (PVs) was able to identify the set of substitutions most likely to have caused a significant shift in PV function during the evolution of *Antarctic notothenioids* in the frigid waters of the Southern Ocean [58]. The results suggest that the current thermal phenotype of Antarctic PVs can be recapitulated by only two amino acid substitutions, namely, K8N and K26N.

These studies were performed by introducing a limited number of mutations into extant proteins, or by carefully selecting ancestors that were separated from an extant protein by only few substitutions. However, such ancestral reconstruction by site-directed mutagenesis appears to be incomplete, as the possibility that sites remaining in a non-ancestral state may significantly affect the molecular property of interest, cannot be ruled out. Although it is difficult and expensive to introduce many mutations into sites widely distributed over gene sequences by site-directed mutagenesis, *de novo* gene synthesis is now available, allowing preparation of ancestral proteins. Therefore, the majority of recent ASR studies have been conducted using full-length or partial ancestral sequence reconstruction, including substitution of corresponding sites in target proteins.

3. Methods for ancestral sequence prediction

How can we determine the sequences of ancestral proteins or genes? In most cases, since the ancestral genes do not currently exist, the ancestral sequences need to be estimated and reconstructed mainly *in silico* (using a computer). Ancestral sequences are calculated using computational methods originally developed for molecular phylogeny construction. Some of these methods, such as maximum parsimony (MP) and maximum likelihood (ML), have an integral procedure of ancestral sequence inference at each node of the phylogenetic tree under construction [65, 66]. The MP method assumes that a phylogenetic tree with minimum substitutions is the most likely. This method assigns a possible nucleotide/amino acid for each site at every node of a phylogenetic tree to evaluate the minimum substitution number. Because of this assumption of parsimony, the MP method tends to underestimate the number of substitutions if a branch is relatively long. The method is also fragile if the evolutionary rate varies among branches.

By contrast, the ML method, which does not require this assumption, is currently more widely used. This method evaluates the posterior probability of a nucleotide/amino acid residue at each node of a phylogenetic tree, based on empirical Bayesian statistics, using the provided sequences and a substitution probability matrix as inputs (observations). Therefore, results can be significantly affected by the choice of input sequences and the choice of substitution probability matrix; the probability of a reconstructed sequence at a node might be low when the node is connected to the provided sequences through longer and/or more intervening branches. The ML method is popular in the field, largely owing to the presence of the excellent software package PAML [67]. Several other software applications have been also developed for this purpose, such as FastML [68], ANSESCON [69], and GASP [70]. With the exception of GASP which partly employs the MP method to enable ancestral state prediction at gapped sites in a sequence alignment, these applications are based on the ML method. In many cases, ancestral sequences cannot be unambiguously determined, and several amino acids might be assigned to a residue site with almost equal probabilities. To avoid false conclusions as a result of such ambiguity, the accuracy of reconstructed ancestral sequence is critical for such studies. However, it is often difficult to obtain a complete, highly accurate sequence, as molecular evolution is believed to be a highly stochastic process and there is no guarantee that ancestral sequences can be identified without errors. Even if each residue of a protein made up of 100 residues, is identified with posterior probability of 0.99 (ie. 99% are expected to be correct), the probability that the sequence as a whole is accurate is only ~0.37 (i.e., 0.99^{100}). In many actual cases, site probabilities are likely to be much lower. This is a major problem in ancestor reconstruction studies, and considerable efforts have been made to avoid incorrect conclusions due to imperfect reconstruction.

Williams et al reported the assessment of the accuracy of ancestral protein reconstruction by MP, ML and Bayesian inference (BI) methods [71]. Their results indicated that MP and ML methods, which reconstruct "best guess" amino acids at each position, overestimate thermostability, while the BI method, which sometimes chooses less-probable residues from the posterior probability distribution, does not. ML and MP tend to eliminate variants at a position that are slightly detrimental to structural stability, simply because such detrimental variants are less frequent. Thus, Williams et al caution that ancestral reconstruction studies require greater care to come to credible conclusions regarding functional evolution [71]. Thornton and colleagues also examined simulation-based experiments, under both simplified and empirically derived conditions, to compare the accuracy of ASR carried out using ML and Bayesian approaches [72]. They showed that incorporating phylogenetic uncertainty by integrating over topologies very rarely changes the inferred ancestral state and does not improve the accuracy of the reconstructed ancestral sequence, suggesting that ML can produce accurate ASRs, even in the face of phylogenetic uncertainty, and using Bayesian integration to incorporate the uncertainty is neither necessary nor beneficial [72].

In the case for experimental molecular archeology using ASR, the effects of equally probable residues at unreliable sites have been tested by site-directed mutagenesis to confirm directly that molecular properties are not largely affected by these. Indeed, in the case of ancestral congerin genes, the single mutant Con-anc'-N28K, in which the suspicious site was replaced with the alternate suggested amino acid was reconstructed in addition to the ancestral congerin (Con-anc', the last common ancestor of ConI and ConII) inferred from the phylogeny of extant galectins using the ML method based on DNA sequences [51]. Nucleotide sequences were retrieved from the DDBJ database [73] , and the ancestral sequence were inferred using the PAML program [67]. The alignment of amino acid sequences of the extant galectins was first prepared using the XCED program [74], and an alignment of the corresponding nucleotide sequences was made in accordance with the amino acid sequence alignment. Tree topology was based on the amino acid sequences of extant proteins using the neighbor-joining (NJ) method. PAML was applied to the phylogeny and alignment to infer the ancestral sequences. The F1X4 matrix was used as the codon substitution model with the universal codon table. The free dN/dS ratio with M8 (beta & omega) model was adapted [75]. The reproduction rate of each Con-anc' amino acid residue was also calculated from the reconstructed sequences, with the exclusion of one extant gene in each case, in order to identify highly unstable sites depending on the choice of extant genes. The results indicated that the average reproduction rate over the sequence was 0.98. The average site posterior probability in the sequence of Con-anc' was 0.81. Seventy-two of 135 sites (53%) had a posterior probability > 0.9. By contrast, 11 sites were found to have posterior probabilities < 0.5. Only one residue, Asn28 of Con-anc', was reproduced with a distinguishably low rate of 0.286, with a suggested alternative amino acid of Lys. Therefore, the single mutant Con-anc'-N28K was also reconstructed. Several reconstruction tests demonstrated that the ancestral sequence had constantly converged into that of Con-anc', and the expected shift by adding a newly found extent sequence was reduced to 1.4% (s.d. 3.2%).

In the case of alcohol dehydrogenase (Adh) ancestral mutants reported by Thomson et al., the posterior probability of the sequence predicted by the ML method was found to be low at three sites. Amino acid residues 168, 211 and 236 of Adh had two (Met and Arg), three (Lys, Arg and Thr), and two (Asp and Asn) equally probable candidates as the ancestral residues, respectively. Therefore, all possible combinations (2 x 3 x 2 = 12) of the candidates at the ambiguous sites were reproduced, and their kinetic properties assessed [13]. The results confirmed with consistency among the alternative mutants that acetaldehyde metabolism was the original function of Adh, that ancestral yeast could not consume ethanol, and that the function of ethanol metabolism was most likely acquired in the linage of the Adh2 locus after gene duplication.

4. Reconstruction of full-length ancestral proteins: Selective adaptive evolution of Conger eel galectins

Conger eel galectins, termed Congerins I and II (Con I and Con II), function as biodefense molecules in the skin mucus and frontier organs including the epidermal club cells of the skin, wall of the oral cavity, pharynx, esophagus, and gills [76-79]. Con I and Con II are prototype galectins, composed of subunits containing 135 and 136 amino acids, respectively, and display 48% amino acid sequence identity [80]. While both Con I and Con II form 2-fold symmetric homodimers with 5- and 6-stranded β-sheets (termed a jellyroll motif), they have different stabilities and carbohydrate-binding specificities, although they do have the conserved carbohydrate recognition domain (CRD) common to other galectins [81-84]. Previous studies of Con I and Con II, based on molecular evolutionary and X-ray crystallography analyses, revealed that these proteins have evolved via accelerated substitutions under natural selection pressure [74-85].

To understand the rapid adaptive differentiation of congerins, experimental molecular archaeology analysis, using the reconstructed ancestral congerins, Con-anc and Con-anc', and their mutants has been conducted [49-51]. Since the ancestral sequences of congerin, Con-anc and Con-anc', were estimated from different phylogenetic trees, which were constructed from the varying numbers of extant genes available (eight for Con-anc, and sixteen for Con-anc') (Fig. 1A), the ancestral sequence Con-anc' showed a 27% discrepancy from the previously inferred sequence of Con-anc (Fig. 1B). Furthermore, as described in the 'Methods for Ancestral Sequence Prediction' section, the reproduction rate of each Con-anc' amino acid residue was examined for the reconstructed sequences, with one extant gene excluded for each estimation, in order to identify highly unstable sites. The result indicated that the average reproduction rate over the sequences were 0.98, and only one residue, Asn28 of Con-anc', was reproduced with a distinguishably low rate of 0.286, prompting verification of the results by the construction of a single mutant Con-anc'-N28K. The revised ancestral congerins, Con-anc' or Con-anc'-N28K, were attached to the nodes of extant proteins with zero distance in the phylogeny constructed from amino acid sequences, indicating that the sequence was appropriate for that of an ancestor (Fig. 1A). On the other hand, the previously inferred Con-anc was attached midway on the ConI branch. Therefore, Con-anc' or Con-anc'-N28K are likely to be closer to the true common ancestor of ConI and ConII than Con-anc. The structures and molecular properties of congerins, as discussed below, also supported this conclusion.

Although Con-anc is an ancestral mutant located midway on the ConI branch and shares a higher sequence similarity with ConI (76%) than with ConII (61%), it showed unique carbohydrate-binding activity and properties, and more closely resembled ConII than ConI, in terms of thermostability and carbohydrate recognition specificity, with the exception of carbohydrates containing α2, 3-sialyl galactose, for example GM3 and GD1a. The ancestral congerins, Con-anc' and Con-anc'-N28K, demonstrated similar carbohydrate binding activity and specificities to those of Con-anc [51]. These analyses of Con-anc suggested a functional evolutionary process for ConI, where it evolved from the ancestral congerin to increase its structural stability and sugar-binding activity. In the case of the ancestral congerin, Con-anc, the candidate amino acid residues responsible for the higher structural stability and carbohydrate-binding activity of Con I were reduced to only 31 amino acid residues, from a total of 71 with apparent differences between Con I and Con II. These were mainly located in the N- and C-terminal and loop regions of the molecule, including the CRD [49, 50]. To identify the residues responsible for the properties of Con I, we next performed molecular evolution tracing analysis, by constructing pseudo-ancestral Con-anc proteins focused on the N-terminal, C-terminal, and some loop regions (loops 3, 5 and 6) [50].

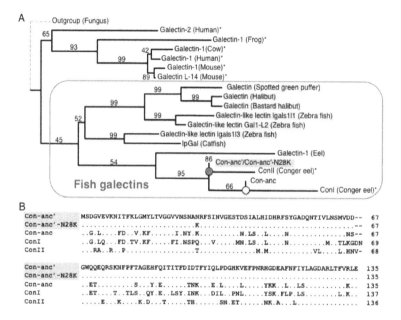

(A) Phylogeny of extant and ancestral congerins. The tree is based on the amino acid sequences of extant galectins and ancestral congerins. The extant genes used for ancestral reconstruction and their accession codes are ConI (*Conger myriaster* congerin I, AB010276.1), ConII (*C. myriaster* congerin II, AB010277.1), *Anguilla japonica* galectin-1 (AJL1, AB098064.1), *Hippoglossus hippoglossus* galectin (AHA1, DQ993254.1), *Paralichthys olivaceus* galectin (PoGal, AF220550.1), *Tetraodon nigroviridis* galectin (TnGal, CR649222.2), *Danio rerio* galectin-like lectin lgals1l1 (DrGal1_L1, BC164225.1), *D. rerio* galectin-like lectin Gal1-L2 (DrGal1_L2, AY421704.1), *D. rerio* Galectin-like lectin lgals1l3 (DrGal1_L3, BC165230.1), *Ictalurus punctatus* galectin (IpGal, CF261531), *Bos taurus* galectin-1 (BTG1, BC103156.1), *Homo sapiens* galectin-1 (HSG1, AK312161.1), *Mus musculus* galectin-1 (MMG1, BC099479.1), *Cricetulus* sp. galectin L-14 (CRG1, M96676.1), *Xenopus laevis* galectin-1 (XLG1, AF170341.1), and *H. sapiens* galectin-2 (HSG2, BC059782.1). The numbers associated with the branches are the percent reproductions of branches in 1000 bootstrap reconstructions. This tree is rooted by using the fungus sequence of *Coprinopsis cinerea* galectin-1 (AF130360.1) as the outgroup. The proteins indicated with asterisk were used for the inference of the previous ancestor (Con-anc). (B) Amino acid sequences of ancestral congerins, ConI, and ConII. Amino acids identical to that of the corresponding last ancestor are represented by a dot.

Figure 1. Amino Acid Sequences and Structures of Ancestral Congerins

This is a protein engineering approach where a proportion of amino acid residues of an extant protein are substituted with those of an ancestor, to construct pseudo-ancestors, in order to reveal the residues determining functional differences between extant and ancestral proteins. These molecular evolutional approaches using pseudo-ancestors bridged from Con-anc to ConI successfully elucidated the regions of the protein relevant to the two adaptive features of ConI, thermostability and higher carbohydrate-binding activity [49]. Experimental molecular archeology analysis, using the reconstructed ancestral congerins, also revealed the process of ConII evolution, another extant galectin. ConII has evolved to enhance affinity for $\alpha 2$, 3-sialyl galactose, which is specifically present in pathogenic marine bacteria. The selection pressure to which Con II reacted was hypothesized to be a shift in carbohydrate affinity. The observed difference in $\alpha 2$, 3-sialyl galactose affinities between Con-anc and Con II support this hypothesis.

The crystal structures of ancestral full-length proteins, Con-anc', Con-anc'-N28K and Con-anc, have been solved at 1.5, 1.6, and 2.0 Å resolutions, respectively [51]. Their three-dimensional (3D) structures clearly demonstrate that Con-anc' or Con-anc'-N28K are appropriate ancestors of extant congerins (Fig. 2). A notable difference between the structures of ConI and ConII is the swapping of S1 strands at the dimer interface, which is unique to ConI among known galectins, and should contribute to its higher stability [81]. The dimer interface of ancestral Con-anc' and Con-anc'-N28K resembled that of ConII, but lacking the strand-swap. This protein-fold is the prototype for dimeric galectins, and the congerin ancestor is expected to have ConII-like conformation. Conversely, Con-anc did display a strand-swapped structure, indicating it was more likely to be an intermediate from the ancestor to ConI, consistent with the results of phylogeny construction (Fig. 2). The differences in carbohydrate interactions between Con-anc' and the extant congerins were observed mainly at the A-face of galactose [51]. These modifications might be relevant to the observed differentiation of carbohydrate specificities between ConI and ConII; ConI prefers $\alpha 1$,4-fucosylated N-acetyl glucosamine, while ConII is adapted to bind $\alpha 2$,3-sialyl galactose-containing carbohydrates [49, 50]. Furthermore, structural or functional parameters, such as cytotoxic activity, thermostability of hemagglutination activity, urea and heat denaturation of the structures, and carbohydrate binding activities of the ancestral and extant congerins, were compared as a function of the evolutionary distances from Con-anc' or ConI [51]. Some of these molecular properties were found to be enhanced in both lineages of congerin, which was observed as a correlation with the evolutionary distance from Con-anc'. The dimer interface essential for these proteins to evoke divalent cross-linking activity was enhanced in both lineages as the number of interface H-bonds and dimer interface area increased in ConI and ConII. However, the lactose interface area and the number of lactose H-bonds showed a low correlation with Ka for carbohydrates, implying that simply enhancing carbohydrate interaction was not likely to be a major selection pressure, and obtaining specificity was more significant for the function of congerins.

Taken together, the first full-length ancestral structures of congerin revealed that the duplicated genes have been differentiating under natural selection pressures for strengthening of the dimer structure and enhancement of the cytotoxic activity. However, the two genes did not react equally to selection pressure, with ConI reacting through protein-fold evolution to enhance its stability. The modification of the dimer interface in the ConII lineage was rather moderate.

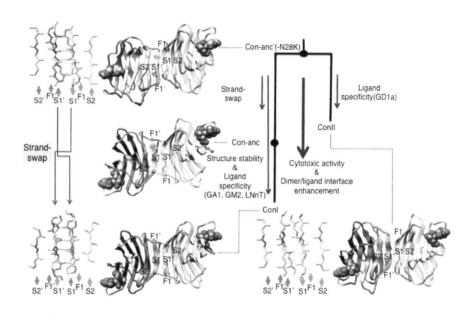

Figure 2. Structures of ConI, ConII, Con-anc and Con-anc'. Con-anc', Con-anc, ConII, and ConI dimers are shown from top to bottom along their molecular phylogeny. The numbers on each branch are the numbers of substitutions. The β strands relevant to the strand-swap at the dimer interface are labeled for S1-S2, and S1'-S2'. Each protein is associated with a close-up of its dimer interface.

5. Reconstruction of ancestral proteins: thermal adaptation of proteins in thermophilic bacterium

Ancestral mutant analysis has been performed to explore the thermal adaptation of proteins. Benner and colleagues reconstructed the ancestral elongation factor-Tu (EF-Tu) predicted using ML methodology, in order to infer the physical environment surrounding ancient organisms [55]. Because EFs play a crucial role in protein synthesis in cells, the thermostability of EFs shows a strong correlation with the optimal growth temperature of their host organisms. For example, the melting temperatures (T_m) of EFs from *Escherichia coli* and *Thermus thermophilus* (HB8) are 42.8°C and 76.7°C, respectively, and the optimal growth temperatures of their respective hosts are approximately 40°C and 74°C, respectively [86]. Thus, EFs are suitable for use in assessment of the ambient temperature at the time of ancient life. To predict the ancestral sequences of EFs, amino acid sequences of fifty EF-Tu proteins from various bacterial lineages were used to construct two kinds of molecular phylogenetic trees; one using the evolutionary distances calculated using the EF-Tu sequences and the second from distances calculated using ribosomal RNA sequences [87]. Both resulting ancestors had temperature profiles similar to that of the thermophilic EF of modern *Thermus aquaticus*, supporting the hypothesis that the common ancestor of all organisms is a hyperthermophile. Inclusion of additional microbial species into the analysis, and reconstruction of the ancestral EFs at various depths (evolutionary distance from present time) in the phylogeny using the ML method [56, 57], demonstrated that ancestral EFs positioned closer to the root of the phylogenetic tree tended to have significantly higher thermostabilities.

Yamagishi and coworkers reported several ancestral proteins, including two metabolic enzymes; 3-isopropylmalate and dehydrogenase (IPMDH), which is involved in leucine biosynthesis, and isocitrate dehydrogenase (ICDH) involved in the TCA cycle. Ancestral amino acids were introduced into extant IPMDH sequences of the hyperthermophilic archaeon *Sulfolobus tokodaii,* the extremely thermophilic bacterium *Thermus thermophilus*, and the hyperthermophilic archaeon *Caldococcus noboribetus* [14-18].

More recently, Hobbs et al reported the reconstruction of several common Precambrian ancestors of the core metabolic enzyme LeuB, 3-isopropylmalate dehydrogenase, estimated from various *Bacillus* species, in addition to the 3D structure of the last common ancestor at 2.9 Å resolution [19]. Their data indicated that the last common ancestor of LeuB was thermophilic, suggesting that the origin of thermophily in the *Bucillus* genus was ancient. Evolutionary tracing analysis through the ancestors of LeuB also indicated that thermophily was not exclusively a primitive trait, and it could be readily gained as well as lost in evolutionary history [19].

Overall, these studies demonstrate that ancestral enzymes retained enzymatic activity and acquired enhanced thermostability over respective extant enzymes, and that introduction of ancestral state amino acids into modern proteins frequently thermostabilizes them. This indicates that ancestral protein reconstruction can provide empirical access to the evolution of ancient phenotypes, and is useful as a strategy for thermostabilization protein engineering.

6. Reconstruction of ancestral proteins: Evolutionary history of nuclear receptors and visual pigment proteins

Thornton and colleagues have reported seminal work using ancestral protein reconstructions of the nuclear receptors for steroid hormones to investigate evolution of their ligand specificities [38-47, 88, 89]. Vertebrates have six homologous nuclear receptors for steroid hormones; the estrogen receptors alpha and beta (ERα and ERβ), androgen receptor (AR), progesterone receptor (PR), glucocorticoid receptor (GR), and mineralocorticoid receptor (MR). As it is thought that these proteins evolved from a common ancestor through a series of gene duplications [65], the reconstruction of their ancestral proteins is a useful tool for investigation of their evolution of ligand-specificity. Although GR and MR are close relatives, GR is activated only by the stress hormone cortisol in most vertebrates, while MR is activated by both aldosterone and cortisol [90, 91]. The amino acid substitutions responsible for the specificity of GR toward cortisol were identified by reconstruction studies of the common ancestor of GRs and MRs using ML methodology [38-47]. Thornton and colleagues also reconstructed the ancestral corticoid receptor (AncCR), which corresponded to the protein predicted to be formed at the duplication event between GR and MR genes. Functional analysis showed that AncCR could be activated by both aldosterone and cortisol, suggesting that GR of vertebrates had lost aldosterone specificity during the evolutionary process. Furthermore, site-direct mutagenesis and X-ray crystallographic studies of AncCR revealed that amino acid substitutions at S106P and L111Q were key for the specificity shift of GR [38, 39]. AncCR is the first complete domain ancestor (ligand-binding domain only), for which 3D structure was determined. Ancestral mutant analysis of the NR5 nuclear orphan receptors, including steroidogenic factor 1 (SF-1) and liver receptor homolog 1 (LRH-1) was also reported [41]. The structure-function relationships of the SF-1/LRH-1 subfamily and their evolutionary ligand-binding shift, where the characteristic phospholipid binding ability of the SF-1/LRH-1 subfamily was subsequently reduced and lost in the lineage leading to the rodent LRH-1, due to specific amino acid replacements, were elucidated [41].

Reconstruction of visual pigment proteins, including rhodopsin and green fluorescent protein (GFP)-like proteins, has been also conducted. Chang et al reconstructed an ancestral archosaur rhodopsin from thirty vertebrate species using the ML method and three different models; nucleotide-, amino acid-, and codon-based [25]. An ancestral protein can be reconstructed with each of these models and the inferred archosaur rhodopsin had the same amino acid sequences for all three, except for three amino acid sites (positions 213, 217, and 218), and all reconstructed ancestral proteins had four variants at the ambiguous sites (single mutants T213I, T217A, V218I, and the triple mutant of these) showed similar optical properties, with an apparent absorption maximum at 508–509 nm, slightly red-shifted from that of modern vertebrates (482–507 nm). These data indicated that the alternative ancestral amino acids predicted by the different likelihood models showed similar functional characteristics. Dim-light and color vision in vertebrates are controlled by five visual pigments (RH1, RH2, SWS1, SWS2, and M/LWS), each consisting of a protein moiety (opsin) and a covalently bound 11-cis-retinal (or 11-cis-3, 4-dehydroretinal) [91], with characteristic sensitivity to specific wavelengths of maximal absorptions (λ_{max}) from 360 nm (UV) to 560 nm (red). How do the visual pigments achieve sensitivity to various wavelengths? Despite extensive mutagenesis analyses of visual pigments, the molecular mechanisms that modulate the variable λ_{max} values observed in nature were not well understood until ancestral protein reconstruction analysis was applied to the question [92]. Yokoyama and colleagues successfully identified the molecular mechanism of the spectral tuning of visual pigments by generating 15 currently known pigment types using engineered ancestral pigments of SWS1, RH1, and red- and green-sensitive (M/LWS) pigments [26-28]. Kawamura and colleagues reported the reconstruction of ancestral mutants of four green visual pigments from zebrafish, namely, RH2-1, RH2-2, RH2-3 and RH2-4, with λ_{max} values of 467, 476, 488, and 505 nm, respectively [29]. The ancestral pigments showed that spectral shifts occurred toward the shorter wavelength in evolutionary lineages [29]. Further-more, Yokoyama and colleagues demonstrated the molecular basis (structural elements) of the adaptation of rhodopsin for the dim-light (deep-sea) vision by ancestral reconstruction experiments using 11 ancestral pigments estimated by rhodopsin sequences of migratory fish from both the surface and deep ocean [30-32].

The great star coral *Montastrea cavernosa* has several green fluorescent protein (GFP)-like proteins, classified into four paralogous groups based on their emission spectra: cyan (emission maximum, 480–495 nm), short wavelength green (500–510 nm), long wavelength green (515–525 nm), and red (575–585 nm) [93]. Matz and colleagues reconstructed the ancestral fluores-cent proteins corresponding to the root of each color group, and the common ancestor of the groups using the ML method [33-37]. The analyses of the fluorescence spectra using the ancestral proteins depicted the evolutionary process of the coral GFP-like proteins, in which the peak wavelength has shifted from green to red. Furthermore, they identified the amino acid substitutions responsible for the generation of recent cyan and red fluorescence proteins through site-direct mutagenesis studies of the ancestral green fluorescent protein as a template [35]. Thus, the engineering of ancestral molecules at various evolutionary stages, to recapitu-late the changes in their phenotypes over time, is an effective way to explore the molecular evolution and adaptation mechanisms of proteins, although the experimental demonstration of adaptive events at the molecular level is particularly challenging.

7. Concluding remarks

Experimental molecular archaeology using ASR is a new and potentially useful method not only for the study of molecular evolution, but also as a protein engineering technique. This method can provide us with experimental information about ancient genes and proteins, which cannot be obtained from fossil records or by simply constructing molecular phylogeny. However, as discussed above, ancestral sequences can have some issues with ambiguity, depending on the choice of evaluation method, evolutionary model, and sequences. Although inference methods such as MP, ML and BI can lead to errors in predicted ancestral sequences, resulting in potentially misleading estimates of the properties of the ancestral protein, experimental molecular archaeology using ASR could be a more reliable method as all possible ancestral mutants, in which ambiguous amino acid sites are replaced by equally probable candidates individually or in combination, are reproducible and the biological and physico-chemical properties and 3D structures of the molecules can be assessed. Indeed, when ancestral congerins were reconstructed based on insufficient sequence information lacking recently determined fish galectin genes, the ancestral Con-anc protein was shown to have a strand-swapped structure resembling ConI, indicating that Con-anc was more likely to be an intermediate mutant of the ancestor to ConI, and that the revised Con-anc' or Con-anc'-N28K are more appropriate ancestors. Thus, the accuracy of ASR can be assessed by analysis of protein activities, stabilities, specificities, and even 3D structures in the laboratory using biochemical or biophysical methods.

Experimental molecular archaeology using ASR can be applied to more complex biological systems, such as heterologous subunit interactions and their evolution in molecular machines [48], host-viral interactions and their co-evolution [54, 94, 95], and proteome/structural proteome level analyses [96]. Furthermore, recent studies have indicated that ASR is applicable to not only to proteins, but also to nucleotides including ancestral rRNA [97] and transposons [95]. To understand the molecular strategies of evolution in nature and the structure-function relationships of proteins and nucleotides, it is important to learn more from 'nature' itself, and from its prodigious works and histories; proteins/nucleotides and their molecular evolution.

Author details

Tomohisa Ogawa[1] and Tsuyoshi Shirai[2]

1 Department of Biomolecular Science, Graduate School of Life Sciences, Tohoku University, Sendai, Japan

2 Nagahama Institute of Bio-Science and Technology, and Japan Science and Technology Agency, Bioinfomatic Research Division, Nagahama, Shiga, Japan

References

[1] Callaway, E. (2010). Fossil genome reveals ancestral link. Nature 468 (7327), 1012.

[2] Pääbo, S, Poinar, H, Serre, D, Jaenicke-després, V, Hebler, J, Rohland, N, Kuch, M, Krause, J, Vigilant, L, & Hofreiter, M. (2004). Genetic analyses from ancient DNA. Ann. Rev. Genet. , 38, 645-679.

[3] Green, R. E, Krause, J, Briggs, A. W, Maricic, T, Stenzel, U, Kircher, M, Patterson, N, Li, H, Zhai, W, Fritz, M. H-Y, Hansen, N. F, Durand, E. Y, Malaspinas, A. S, Jensen, J. D, Marques-bonet, T, Alkan, C, Prüfer, K, Meyer, M, Burbano, H. A, Good, J. M, Schultz, R, Aximu-petri, A, Butthof, A, Höber, B, Höffner, B, Siegemund, M, Weih-mann, A, Nusbaum, C, Lander, E. S, Russ, C, Novod, N, Affourtit, J, Egholm, M, Ver-na, C, Rudan, P, Brajkovic, D, Kucan, Ž, Gušic, I, Doronichev, V. B, Golovanova, L. V, Lalueza-fox, C, De La Rasilla, M, Fortea, J, Rosas, A, & Schmitz, R. W. Johnson PLF, Eichler EE, Falush D, Birney E, Mullikin JC, Slatkin M, Nielsen R, Kelso J, Lachmann M, Reich D, Pääbo S. ((2010). A draft sequence of the neandertal genome. Science 328 (5979), 7710-7722.

[4] Krause, J, Fu, Q, Good, J. M, Viola, B, Shunkov, M. V, Derevianko, A. P, & Pääbo, S. The complete mitochondrial DNA genome of an unknown hominin from southern Siberia. ((2010). Nature, 464 (7290), 894-897.

[5] Reich, D, Green, R. E, Kircher, M, Krause, J, Patterson, N, Durand, E. Y, Viola, B, Briggs, A. W, & Stenzel, U. Johnson PLF, Maricic T, Good JM, Marques-Bonet T, Alkan C, Fu Q, Mallick S, Li H, Meyer M, Eichler EE, Stoneking M, Richards M, Talamo S, Shunkov MV, Derevianko AP, Hublin J-J, Kelso J, Slatkin M, Pääbo S. ((2010). Genetic history of an archaic hominin group from Denisova cave in Siberia. Nature 468 (7327), 1053-1060.

[6] Rohland, N, Reich, D, Mallick, S, Meyer, M, Green, R. E, Georgiadis, N. J, Roca, A. L, & Hofreiter, M. (2010). Genomic DNA sequences from mastodon and woolly mammoth reveal deep speciation of forest and savanna elephants. PLoS biol., 8 (12), e1000564

[7] Gibson, D. G, Glass, J. I, Lartigue, C, Noskov, V. N, Chuang, R-Y, Algire, M. A, Benders, G. A, Montague, M. G, Ma, L, Moodie, M. M, Merryman, C, Vashee, S, Krishnakumar, R, Assad-garcia, N, Andrews-pfannkoch, C, Denisova, E. A, Young, L, Qi, Z-Q, Segall-shapiro, T. H, Calvey, C. H, & Parmar, P. P. Hutchison III CA, Smith HO, Venter JC. ((2011). Creation of a Bacterial Cell Controlled by a Chemically Synthesized Genome Science 329 (5987), 52-56.

[8] Lai, J, Jin, J, Kubelka, J, & Liberles, D. A. (2012). A Phylogenetic Analysis of Normal Modes Evolution in Enzymes and Its Relationship to Enzyme Function. J. Mol. Biol. , 422, 442-459.

[9] Harms, M. J, & Thornton, J. W. (2010). Analyzing protein structure and function using ancestral gene reconstruction. Curr. Opin. Struct. Biol. 20 (3), 360-366.

[10] Stackhouse, J, Presnell, S. R, Mcgeehan, G. M, Nambiar, K. P, & Benner, S. A. (1990). The ribonuclease from an extinct bovid ruminant. FEBS Lett., , 262, 104-106.

[11] Jermann, T. M, Opitz, J. G, Stackhouse, J, & Benner, S. A. (1995). Reconstructing the evolutionary history of the artiodactyl ribonuclease superfamily. Nature , 374, 57-59.

[12] Malcolm, B. A, Wilson, K. P, Matthews, B. W, Kirsch, J. F, & Wilson, A. C. (1990). Ancestral lysozymes reconstructed, neutrality tested, and thermostability linked to hydrocarbon packing. Nature , 345, 86-89.

[13] Thomson, J. M, Gaucher, E. A, Burgan, M. F, De Kee, D. W, Li, T, Aris, J. P, & Benner, S. A. (2005). Resurrecting ancestral alcohol dehydrogenases from yeast. Nat. Genet. , 37, 630-635.

[14] Miyazaki, J, Nakaya, S, Suzuki, T, Tamakoshi, M, Oshima, T, & Yamagishi, A. (2001). Ancestral residues stabilizing isopropylmalate dehydrogenase of an extreme thermophile: Experimental evidence supporting the thermophilic common ancestor hypothesis. J. Biochem. 129 (5), 777-782., 3.

[15] Watanabe, K, Ohkuri, T, Yokobori, S, & Yamagishi, A. (2006). Designing thermostable proteins: ancestral mutants of 3-isopropylmalate dehydrogenase designed by using a phylogenetic tree. J. Mol. Biol. , 355, 664-674.

[16] Watanabe, K, & Yamagishi, A. (2006). The effects of multiple ancestral residues on the Thermus thermophilus isopropylmalate dehydrogenase. FEBS Lett. 580 (16), 3867-3871., 3.

[17] Dean, A. M, & Golding, G. B. (1997). Protein engineering reveals ancient adaptive replacements in isocitrate dehydrogenase. Proc. Natl. Acad. Sci. USA 94 (7), 3104-3109.

[18] Iwabata, H, Watanabe, K, Ohkuri, T, Yokobori, S, & Yamagishi, A. (2005). Thermostability of ancestral mutants of Caldococcus noboribetus isocitrate dehydrogenase. FEMS Microbiol. Lett., , 243, 393-398.

[19] Hobbs, J. K, Shepherd, C, Saul, D. J, Demetras, N. J, Haaning, S, Monk, C. R, Daniel, R. M, & Arcus, V. L. (2012). On the origin and evolution of thermophily: Reconstruction of functional precambrian enzymes from ancestors of Bacillus. Mol. Biol. Evol. 29 (2), 825-835.

[20] Chandrasekharan, U. M, Sanker, S, Glynias, M. J, Karnik, S. S, & Husain, A. (1996). Angiotensin II-forming activity in a reconstructed ancestral chymase. Science , 271, 502-505.

[21] Akanuma, S, Iwami, S, Yokoi, T, Nakamura, N, Watanabe, H, Yokobori, S-I, & Yamagishi, A. (2011). Phylogeny-based design of a B-subunit of DNA gyrase and its ATPase domain using a small set of homologous amino acid sequences. J. Mol. Biol. 412 (2), 212-225.

[22] Shimizu, H, Yokobori, S-i, Ohkuri, T, Yokogawa, T, Nishikawa, K, & Yamagishi, A. (2007). Extremely Thermophilic Translation System in the Common Ancestor Commonote: Ancestral Mutants of Glycyl-tRNA Synthetase from the Extreme Thermophile Thermus thermophiles. J. Mol. Biol. 369 (4), 1060-1069.

[23] Alcolombri, U, Elias, M, & Tawfik, D. S. (2011). Directed evolution of sulfotransferases and paraoxonases by ancestral libraries. J. Mol. Biol. 411 (4), 837-853.

[24] Perez-jimenez, R, Inglés-prieto, A, Zhao, Z-M, Sanchez-romero, I, Alegre-cebollada, J, Kosuri, P, Garcia-manyes, S, Kappock, T. J, Tanokura, M, Holmgren, A, Sanchez-ruiz, J. M, Gaucher, E. A, & Fernandez, J. M. (2011). Single-molecule paleoenzymology probes the chemistry of resurrected enzymes. Nature Struct. Mol. Biol. 18 (5), 592-596.

[25] Chang, B. S, Jonsson, K, Kazmi, M. A, Donoghue, M. J, & Sakmar, T. P. (2002). Recreating a functional ancestral archosaur visual pigment. Mol. Biol. Evol. , 19, 1483-1489.

[26] Shi, Y, & Yokoyama, S. (2003). Molecular analysis of the evolutionary significance of ultraviolet vision in vertebrates. Proc. Natl. Acad. Sci. USA. , 100, 8308-8313.

[27] Yokoyama, S, & Takenaka, N. (2004). The molecular basis of adaptive evolution of squirrelfish rhodopsins. Mol. Biol. Evol. , 21, 2071-2078.

[28] Yokoyama, S, Yang, H, & Starmer, W. T. (2008). Molecular basis of spectral tuning in the red- and green-sensitive (M/LWS) pigments in vertebrates. Genetics , 179, 2037-2043.

[29] Chinen, A, Matsumoto, Y, & Kawamura, S. (2005). Reconstitution of ancestral green visual pigments of zebrafish and molecular mechanism of their spectral differentiation. Mol. Biol. Evol., , 22, 1001-1010.

[30] Yokoyama, S, Tada, T, Zhang, H, & Britt, L. (2008). Elucidation of phenotypic adaptations: Molecular analyses of dim-light vision proteins in vertebrates. Proc. Natl. Acad. Sci. USA., , 105, 13480-13485.

[31] Yokoyama, S. (2008). Evolution of dim-light and color vision pigments. Annu. Rev. Genom. Hum Genet., , 9, 259-282.

[32] Watanabe, H. C, Mori, Y, Tada, T, Yokoyama, S, & Yamato, T. (2010). Molecular mechanism of long-range synergetic color tuning between multiple amino acid residues in conger rhodopsin. Biophysics , 6, 67-78.

[33] Ugalde, J. A, Chang, B. S, & Matz, M. V. (2004). Evolution of coral pigments recreated. Science, 305, 1433.

[34] Chang BSWUgalde JA, Matz MV. ((2005). Applications of ancestral protein reconstruction in understanding protein function: GFP-like proteins. Method Enzymol. , 395, 652-670.

[35] Field, S. F, Bulina, M. Y, Kelmanson, I. V, Bielawski, J. P, & Matz, M. V. (2006). Adaptive evolution of multicolored fluorescent proteins in reef-building corals. J. Mol. Evol. 62 (3), 332-339.

[36] Alieva, N. O, Konzen, K. A, Field, S. F, Meleshkevitch, E. A, Hunt, M. E, Beltran-ramirez, V, Miller, D. J, Wiedenmann, J, Salih, A, & Matz, M. V. (2008). Diversity and evolution of coral fluorescent proteins. PLoS ONE 3 (7), art. (e2680)

[37] Field, S. F, & Matz, M. V. (2010). Retracing evolution of red fluorescence in GFP-like proteins from faviina corals. Mol. Biol. Evol. 27 (2), 225-233.

[38] Thornton, J. W. (2001). Evolution of vertebrate steroid receptors from an ancestral estrogen receptor by ligand exploitation and serial genome expansions. Proc. Natl. Acad. Sci. USA, , 98, 5671-5676.

[39] Thornton, J. W, Need, E, & Crews, D. (2003). Resurrecting the ancestral steroid receptor: Ancient origin of estrogen signaling. Science 301 (5640), 1714-1717.

[40] Thornton, J. W. (2004). Resurrecting ancient genes: Experimental analysis of extinct molecules. Nat. Rev. Genet. 5 (5), 366-375.

[41] Krylova, I. N, Sablin, E. P, Moore, J, Xu, R. X, & Waitt, G. M. MacKay JA, Juzumiene D, Bynum JM, Madauss K, Montana V, Lebedeva L, Suzawa M, Williams JD, Williams SP, Guy RK, Thornton JW, Fletterick RJ, Willson TM, Ingraham HA. ((2005).

Structural analyses reveal phosphatidyl inositols as ligands for the NR5 orphan receptors SF-1 and LRH-1. Cell 120 (3), 343-355.

[42] Bridgham, J. T, Carroll, S. M, & Thornton, J. W. (2006). Evolution of hormone-receptor complexity by molecular exploitation. Science 312 (5770), 97-101.

[43] Ortlund, E. A, Bridgham, J. T, Redinbo, M. R, & Thornton, J. W. (2007). Crystal structure of an ancient protein: Evolution by conformational epistasis. Science 317 (5844), 1544-1548.

[44] Bridgham, J. T, Ortlund, E. A, & Thornton, J. W. (2009). An epistatic ratchet constrains the direction of glucocorticoid receptor evolution. Nature 461 (7263), 515-519

[45] Bridgham, J. T, Eick, G. N, Larroux, C, Deshpande, K, & Harms, M. J. Gauthier MEA, Ortlund EA, Degnan BM, Thornton JW. ((2010). Protein evolution by molecular tinkering: Diversification of the nuclear receptor superfamily from a ligand-dependent ancestor. PLoS Biology 8 (10), art. (e1000497)

[46] Eick, G. N, & Thornton, J. W. (2011). Evolution of steroid receptors from an estrogen-sensitive ancestral receptor. Mol. Cell. Endocrinol. 334 (1-2), 31-38.

[47] Carroll, S. M, Ortlund, E. A, & Thornton, J. W. (2011). Mechanisms for the evolution of a derived function in the ancestral glucocorticoid receptor. PLoS Genetics 7 (6), art. (e1002117)

[48] Finnigan, G. C, Hanson-smith, V, Stevens, T. H, & Thornton, J. W. (2012). Evolution of increased complexity in a molecular machine. Nature 481 (7381), 360-364.

[49] Konno, A, Ogawa, T, Shirai, T, & Muramoto, K. (2007). Reconstruction of a probable ancestral form of conger eel galectins revealed their rapid adaptive evolution process for specific carbohydrate recognition. Mol. Biol. Evol., , 24, 2504-2514.

[50] Konno, A, Yonemaru, S, Kitagawa, A, Muramoto, K, Shirai, T, & Ogawa, T. (2010). Protein engineering of conger eel galectins by tracing of molecular evolution using probable ancestral mutants. BMC Evol. Biol., 10:43, doi:10.1186/1471-2148-10-43.

[51] Konno, A, Kitagawa, A, Watanabe, M, Ogawa, T, & Shirai, T. (2011). Tracing protein evolution through ancestral structures of fish galectin.
Structure 19 (5), 711-721.

[52] Yadid, I, & Tawfik, D. S. (2011). Functional-propeller lectins by tandem duplications of repetitive units. Protein Eng. Design Select. 24 (1-2), 185-195.

[53] Gullberg, M, Tolf, C, Jonsson, N, Mulders, M. N, Savolainen-kopra, C, Hovi, T, Van Ranst, M, Lemey, P, Hafenstein, S, & Lindberg, A. M. Characterization of a putative ancestor of coxsackievirus B5. J. Virology 84 (19), 9695-9708.

[54] Kaiser, S. M, Malik, H. S, & Emerman, M. (2007). Restriction of an extinct retrovirus by the human TRIM5α antiviral protein. Science , 316, 1756-1758.

[55] Gaucher, E. A, Thomson, J. M, Burgan, M. F, & Benner, S. A. (2003). Inferring the pa-
laeoenvironment of ancient bacteria on the basis of resurrected proteins. Nature 425
(6955), 285-288.

[56] Gaucher, E. A, Govindarajan, S, & Ganesh, O. K. (2008). Palaeotemperature trend for
Precambrian life inferred from resurrected proteins. Nature 451 (7179), 704-707.

[57] Gouy, M, & Chaussidon, M. (2008). Evolutionary biology: ancient bacteria liked it
hot. Nature, , 451, 635-636.

[58] Whittington, A. C, & Moerland, T. S. (2012). Resurrecting prehistoric parvalbumins
to explore the evolution of thermal compensation in extant Antarctic fish parvalbumins. J. Exp. Biol. 215 (18), 3281-3292.

[59] Hult, E. F, Weadick, C. J, Chang, B. S. W, & Tobe, S. S. (2008). Reconstruction of ancestral FGLamide-type insect allatostatins: A novel approach to the study of allatos-
tatin function and evolution. J. Insect Physiol. 54 (6), 959-968.

[60] Skovgaard, M, Kodra, J. T, Gram, D. X, Knudsen, S. M, Madsen, D, & Liberles, D. A.
(2008). Using Evolutionary Information and Ancestral Sequences to Understand the
Sequence-Function Relationship in GLP-1 Agonists J.Mol. Biol. 363 (5), 977-988.

[61] Beintema, J. J. (1987). Structure, properties and molecular evolution of pancreatic-
type ribonucleases. Life Chem. Rep., , 4, 333-389.

[62] Carroll, R. L. (1988). Vertebrate Paleontology and Evolution, New York, Freeman.

[63] Woese, C. R. (1987). Bacterial evolution. Microbiol. Rev., , 51, 221-271.

[64] Imada, K, Sato, M, Tanaka, N, Katsube, Y, Matsuura, Y, & Oshima, T. (1991). Three-
dimensional structure of a highly thermostable enzyme, 3-isopropylmalate dehydrogenase of Thermus thermophilus at 2.2 A resolution. J. Mol. Biol., , 222, 725-738.

[65] Eck, R. V, & Dayhoff, M. O. (1966). Evolution of the structure of ferredoxin based on
living relics of primitive amino acid sequences. Science, , 152, 363-366.

[66] Felsenstein, J. (1981). Evolutionary trees from DNA sequences: a maximum likelihood approach. J. Mol. Evol., , 17, 368-376.

[67] Yang, Z. (2007). PAML 4: phylogenetic analysis by maximum likelihood. Mol. Biol.
Evol., , 24, 1586-1591.

[68] Pupko, T. Pe'er, I., Shamir, R. & Graur, D. ((2000). A fast algorithm for joint reconstruction of ancestral amino acid sequences. Mol. Biol. Evol. , 17, 890-896.

[69] Cai, W, Pei, J, & Grishin, N. V. (2004). Reconstruction of ancestral protein sequences
and its applications. BMC Evol. Biol., 4, 33.

[70] Edwards, R. J, & Shields, D. C. (2004). GASP: Gapped ancestral sequence prediction
for proteins. BMC Bioinf., 5, 123.

[71] Williams, P. D, Pollock, D. D, Blackburne, B. P, & Goldstein, R. A. (2006). Assessing the accuracy of ancestral protein reconstruction methods. PLoS Comput. Biol. 2 (6), 0598-0605.

[72] Hanson-smith, V, Kolaczkowski, B, & Thornton, J. W. (2010). Robustness of ancestral sequence reconstruction to phylogenetic uncertainty. Mol. Biol. Evol. 27 (9), 1988-1999.

[73] Kaminuma, E, Mashima, J, Kodama, Y, Gojobori, T, Ogasawara, O, Okubo, K, Takagi, T, & Nakamura, Y. (2010). DDBJ launches a new archive database with analytical tools for next-generation sequence data. Nucleic Acids Res. 38, DD38., 33.

[74] Katoh, K, Misawa, K, Kuma, K, & Miyata, T. (2002). MAFFT: a novel method for rapid multiple sequence alignment based on fast Fourier transform. Nucleic Acids Res. , 30, 3059-3066.

[75] Yang, Z, Nielsen, R, Goldman, N, & Pedersen, A. M. (2000). Codon-substitution models for heterogeneous selection pressure at amino acid sites. Genetics , 155, 431-449.

[76] Kamiya, H, Muramoto, K, & Goto, R. (1988). Purification and properties of agglutinins from conger eel, Conger myriaster (Brevoort), skin mucus. Dev. Comp. Immuno., , 12, 309-318.

[77] Muramoto, K, & Kamiya, H. (1992). The amino-acid sequence of a lectin from conger eel, Conger myriaster, skin mucus. Biochem. Biophys. Acta, , 1116, 129-136.

[78] Muramoto, K, Kagawa, D, Sato, T, Ogawa, T, Nishida, Y, & Kamiya, H. (1999). Functional and structural characterization of multiple galectins from the skin mucus of conger eel, Conger myriaster. Comp. Biochem. Physiol. B, , 123, 33-45.

[79] Nakamura, O, Matsuoka, H, Ogawa, T, Muramoto, K, Kamiya, H, & Watanabe, T. (2006). Opsonic effect of congerin, a mucosal galectin of the Japanese conger, Conger myriaster (Brevoort). Fish Shellfish Immunol. , 20, 433-435.

[80] Ogawa, T, Ishii, C, Kagawa, D, Muramoto, K, & Kamiya, H. (1999). Accelerated evolution in the protein-coding region of galectin cDNAs, congerin I and congerin II, from skin mucus of conger eel (Conger myriaster). Biosci. Biotechnol. Biochem., , 63, 1203-1208.

[81] Ogawa, T, Shirai, T, Shionyu-mitsuyama, C, Yamane, T, Kamiya, H, & Muramoto, K. (2004). The speciation of conger eel galectins by rapid adaptive evolution. Glycoconj. J., , 19, 451-458.

[82] Shirai, T, Mitsuyama, C, Niwa, Y, Matsui, Y, Hotta, H, Yamane, T, Kamiya, H, Ishii, C, Ogawa, T, & Muramoto, K. (1999). High-resolution structure of the conger eel galectin, congerin I, in lactose-liganded and ligand-free forms: emergence of a new structure class by accelerated evolution. Structure, , 7, 1223-1233.

[83] Shirai, T, Matsui, Y, Shionyu-mitsuyama, C, Yamane, T, Kamiya, H, Ishii, C, Ogawa, T, & Muramoto, K. (2002). Crystal structure of a conger eel galectin (Congerin II) at

1.45 angstrom resolution: Implication for the accelerated evolution of a new ligand-binding site following gene duplication. J. Mol. Biol., , 321, 879-889.

[84] Shirai, T, Shionyu-mitsuyama, C, Ogawa, T, & Muramoto, K. (2006). Structure based studies of the adaptive diversification process of congerins. Mol. Div., , 10, 567-573.

[85] Ogawa, T. (2006). Molecular diversity of proteins in biological offense and defense systems. Mol. Div., , 10, 511-514.

[86] Williams, R. A. D. da Costa, M. S. ((1992). The genus Thermus and related microorganisms. In Balows, A. Truper, H. G., Dworkin, M., Harder, W. & Schleifer, K.-H. (Eds.), The Prokaryotes (2nd edition, New York, Springer., 3745-3753.

[87] Hugenholtz, P, Goebel, B. M, & Pace, N. R. (1998). Impact of culture-independent studies on the emerging phylogenetic view of bacterial diversity. J. Bacteriol., , 180, 4765-4774.

[88] Thornton, J. W, & Desalle, R. (2000). A new method to localize and test the significance of incongruence: detecting domain shuffling in the nuclear receptor superfamily. Syst. Biol., , 49, 183-201.

[89] Dean, A. M, & Thornton, J. W. (2007). Mechanistic approaches to the study of evolution: The functional synthesis. Nat. Rev. Genet. 8(9), 675-688.

[90] Bentley, P. J. (1998). Comparative Vertebrate Endocrinology, Cambridge Univ. Press, Cambridge.

[91] Farman, N, & Rafestin-oblin, M. E. (2001). Multiple aspects of mineralocorticoid selectivity. Am. J. Physiol. Renal. Physiol., 280. F , 181-192.

[92] Yokoyama, S. (2000). Molecular evolution of vertebrate visual pigments. Prog. Retin. Eye Res. , 19, 385-419.

[93] Kelmanson, I. V, & Matz, M. V. (2003). Molecular basis and evolutionary origins of color diversity in great star coral Montastraea cavernosa (Scleractinia: Faviida). Mol. Biol. Evol., , 20, 1125-1133.

[94] Heinemann, J, Maaty, W. S, Gauss, G. H, Akkaladevi, N, Brumfield, S. K, Rayaprolu, V, Young, M. J, Lawrence, C. M, & Bothner, B. (2011). Fossil record of an archaeal HK97-like provirus. Virology , 417, 362-368.

[95] Münk, C, Willemsen, A, & Bravo, I. G. (2012). An ancient history of gene duplications, fusions andlosses in the evolution of APOBEC3 mutators inmammals. BMC Evol. Biol., 12, 71 doi:10.1186/1471-2148-12-71.

[96] Erdin, S, Ward, R. M, Venner, E, & Lichtarge, O. (2010). Evolutionary Trace Annotation of Protein Function in the Structural Proteome. J. Mol. Biol. , 396, 1451-1473.

[97] Lu, Q, & Fox, G. E. Resurrection of an ancestral 5S rRNA ((2011). BMC Evol. Biol. 11 (1), art. (218)

Chapter 6

Identification of HMGB1-Binding Components Using Affinity Column Chromatography

Ari Rouhiainen, Helena Tukiainen, Pia Siljander and
Heikki Rauvala

Additional information is available at the end of the chapter

1. Introduction

High Mobility Group B1 (HMGB1) is a 30 kDa protein widely expressed in mammalian cells. HMGB1 has a high content of charged amino acids and has a bipolar structure consisting of two highly positive amino terminal HMG-box domains and an acidic carboxy terminal tail. HMGB1 has nuclear functions regulating chromatin structure and gene expression and extracellular functions regulating immune response and cell motility. Biochemical and cell biological studies have revealed that HMGB1 binds to various kinds of biomolecules and these interactions are crucial for determining the *in vivo* functions of HMGB1. Albeit several different biochemical methods have been used to detect HMGB1-binding components, HMGB1-affinity column chromatography has rarely been applied in such studies. Here, we describe an affinity chromatography method that we have applied to isolation and identification of HMGB1-binding molecules from different cell types. Biomolecules recovered with HMGB1-affinity chromatography include proinflammatory bacterial DNA and glioblastoma cell histones H1 and H3 which all have previously been reported as HMGB1-binding molecules by other methods. Furthermore, an entirely new HMGB1-binding protein, Multimerin-1 containing complex, was identified from platelet lysates by HMGB1-affinity chromatography. Endogenous Multimerin-1 and HMGB1 were shown to associate on the surface of endothelial cells and activated platelets, and endogenous Multimerin-1 also regulated the release of HMGB1 from activated platelets. In conclusion, HMGB1-affinity chromatography can be used to isolate and characterize novel HMGB1-binding partners from a variety of cellular sources. Such new interactions reveal further complexity in the multi-faceted biology of the HMGB1.

2. HMGB1

2.1. HMGB1 (Amphoterin) as a heparin-binding protein

Extracellular form of HMGB1 was originally identified from the developing rat brain as an adhesive neurite outgrowth promoting molecule. This finding was done in studies where brain lysates were fractionated with heparin affinity chromatography and the activity of brain tissue fractions to induce neurite outgrowth was monitored [1]. Since the isolated neurite outgrowth promoting protein had a bi-polar structure it was named as "Amphoterin" [2]. Its structure turned out to be identical with a previously characterised nuclear DNA-binding protein "High Mobility Group -1 [3]. In current nomenclature this protein is called HMGB1 [4].

In addition to heparin-Sepharose affinity chromatography HMGB1 has been captured as a ligand in other affinity column chromatographies. These studies have been performed using chromatographies where proteins, nucleic acids, lipids or carbohydrates have been used as baits coupled to solid matrices. Studies where HMGB1 has specifically been shown to bind to novel ligands include the use of Receptor for Advanced Glycation End Products (RAGE) [5], single stranded DNA [6], sulfatide [7] and carboxylated glycan [8] affinity columns.

2.2. HMGB1-column affinity chromatography

HMGB1 consists of three domains. Two of them, Box A and Box B, are DNA-binding domains, while the third is an acidic carboxy terminal domain that binds histones H1 and H3 [reviewed in 9 and 10]. The use of HMGB1 affinity column chromatographies in isolation of proteins has been described in the literature. HMGB1-Sepharose column was used as a negative control in a study where a specific Syndecan-3 binding partner was detected. Although Syndecan-3 has highly sulphated glycosaminoglycan side chains, it did not bind to the HMGB1-Sepharose [11]. HMGB1-domain affinity column chromatography has also been used to study other HMGB1 interactions. Thus, Hox proteins were discovered as HMGB1-binding proteins using HMGB-domain coupled Sepharose column chromatography [12]. In addition, the HMGB1-binding peptide of RAGE was determined using HMGB1-domain coupled -Sepharose chromatography [13].

2.3. HMGB1 as a modulator of innate immunity

Most of the HMGB1 studies performed during the last decade have been focused on its functions as a modulator of inflammation. The role of HMGB1 in the immune system was recognised more than a decade ago [14]. Since then additional studies have confirmed that elevated tissue levels of HMGB1 can serve as a general marker of inflammation or tissue damage [reviewed in 15,16 and 17].

In blood circulation, leukocytes [18], platelets [19] and endothelial cells [20] express HMGB1. In unactivated nuclear cells, HMGB1 localises to the nucleus, and in resting platelets, HMGB1 localises to the cytoplasm. After activation, HMGB1 localises towards the periphery of the cell and it is released to the extracellular space via an unconventional secretion pathway [18, 21]. Recently, HMGB1 was also detected in platelet-derived microvesicles [22]. As a result, serum HMGB1-levels are elevated and the level of HMGB1 in the serum often correlates to the disease severity [23]. Further, the extent of the post translational modifications (acetylation of lysines, proteolytic cleavage or oxidation of cysteines) of HMGB1 correlates to various pathological states [24, 25, 26].

2.4. Platelets

Platelets mediate haemostasis when vessel wall is injured. During hemostasis, platelet receptors work in sequence by slowing down platelets via glycoprotein (GP) Ib/von Wille-brand factor to bring them into contact with subendothelial matrix proteins e.g. collagen, then activating them via GPVI to release the contents of various granules and to express integrins in an active state. Thrombus forms when more platelets become incorporated via activated integrin αIIbβ3, which is bridged by several adhesive RGD-containing proteins, but other receptors also finetune the response. Incoming platelets are incorporated into the growing thrombus, until eventually, this activity decreases limiting the size of the thrombus. Simulta-neously to the adhesive process, platelets engage in the procoagulant transformation and thrombin generation by changes in the plasma membrane phospholipids and by liberating coagulation factors, e.g. factor V/Va. Platelets also release microvesicles which may spread activation-promoting molecules or recruit heterogenic cell interactions. The final result is that the exposed subendothelium becomes protected by a non-thrombogenic platelet surface while maintaining blood flow. Subsequently, events of tissue repair are initiated to restore the vascular wall and finally to remove the thrombus. In contrast to normal hemostasis, patho-logical thrombosis manifests these events in an uncontrolled fashion yielding to cessation of blood flow.

Platelets have also recently been discovered to play a role in immunity [27], development [28] and neuroinflammation [29]. Therefore, various heterogenic cell interactions of platelets with other cells such as leukocytes or endothelial cells are of current research interest. Platelets have several interactive mechanisms ranging from the liberation of bioactive molecules to direct membrane-borne cell-cell interactions (e.g. CD62P, CD40L, JAM-A, GPIb) [30].

Platelets contain and liberate bioactive molecules including proteins, lipids and even mRNA and miRNA, which allow them to participate in and modulates diverse physiological func-tions, some of which at the first glance may seem to be contradictory e.g., pro-and anticoagulant functions. A proteomic analysis of the releasate from TRAP-activated platelets showed that in addition to the nearly 400 previously identified α-granule- or microvesicle-associated mole-cules, platelets liberated over 300 previously unrecognized molecules [31]. Many of these proteins have interactions with each other which further modulate their biological functions. Clearly, dynamics, mechanisms and selectivity of the different secretion processes must be carefully controlled in the platelets. Versatility seems to be a key property of the platelet.

Platelets bind to HMGB1 but the cell surface receptor mediating this interaction is unknown. Platelets express previously recognised HMGB1 receptors TLR2/4/9 [27], RAGE [32], trans-membrane proteoglycans [33] and anionic lipids [19]. Whether these structures mediate HMGB1 binding to platelets has not been much studied. However, previously one study showed that RAGE and TLR2 could mediate HMGB1 binding to platelets [34]. Inhibition of coagulation or platelet activation *in vivo* reduced serum HMGB1-levels in rat disease models suggesting that platelets are an important source of circulating HMGB1 [35, 36]. Further, circulating HMGB1 levels were shown to correlate with platelet activation markers in patients with hematologic malignancy and disseminated intravascular coagulation [37].

3. Materials and methods

3.1. HMGB1-column affinity chromatographies

3.1.1. Binding of bacterial DNA to HMGB1 affinity column

Recombinant HMGB1 was coupled to activated Sepharose 1 ml High-Trap –column (GE Healthcare, Little Chalfont, UK) according to manufactur´s protocols [38]. HMGB1-affinity chromatography of bacterial lysates and the quantitative analysis of eluted DNA were conducted as described [38].

3.1.2. Binding of histones H1 and H3 to HMGB1 affinity column

C6 glioblastoma cell [39] nuclear fraction was isolated as described [40] and analysed in HMGB1-Sepharose affinity column (1 ml High-Trap column) chromatography using ÄKTA Micro High-performance liquid chromatography station (GE Healthcare) with phosphate buffered saline (PBS) as a chromatography buffer. Flow rate in chromatography was 1 ml/min. HMGB1 column bound proteins were eluted using linear NaCl-gradient (0.15-2 M NaCl, 20 min). 0.5 ml fractions were collected. The protein-containing fractions were analysed in a histone ELISA. Briefly, MaxiSorp microwell plates (Nunc, Roskilde, Denmark) were coated with 2 µg/µl of anti-PAN-Histone antibody (Millipore Corporation, Billerica, MA, USA) in PBS overnight at 4 °C, and wells were blocked with bovine serum albumin (BSA, Sigma-Aldrich, St. Louis, MO, USA). HMGB1 column eluted C6 cell nuclear protein fractions were diluted seven-fold with 10 mM Tris, pH 7.5, and samples were applied to the wells. After 1h incubation at 37 °C, wells were washed with PBS and the bound histones H1 and H3 were detected with 1/1000 dilution of rabbit anti-Histone H1 antibody (Active Motif, Carlsbad, CA, USA) or rabbit anti-Histone H3 antibody (Cell Signaling Technology, Danvers, MA, USA), followed by horseradish peroxidase -conjugated anti-Rabbit IgG (GE Healthcare) detection..

3.1.3. Isolation of HMGB1 binding component from platelets

Outdated platelet concentrates (Finnish Red Cross Blood Service, Helsinki, Finland) were lysed in lysis buffer [10 mM Tris-HCl, pH 8.5 containing protease inhibitor cocktail (Roche, Basel, Switzerland)] and centrifuged for 2 h at 20 000 g (4 °C). The pellet was extracted overnight at 4 °C with 50 mM octyl-glucoside (Sigma-Aldrich) – PBS containing 10 µg/ml of aprotinin (Sigma-Aldrich) and 0.1 mM phenylmethylsulfonyl fluoride (Sigma-Aldrich). The homogenate was centrifuged for 2 h at 20 000 g and the supernatant was diluted 10 times with PBS and applied to diethylaminoethanol-Sepharose –column (GE Healthcare). The column was washed with 5 mM octyl-glucoside – PBS containing 10 µg/ml aprotinin and 0.1 mM phenyl-methylsulfonyl fluoride and bound proteins were eluted with 2 M NaCl 5 mM octyl-glucoside – PBS containing 10 µg/ml aprotinin and 0.1 mM phenylmethylsulfonyl fluoride. HMGB1-binding fractions were identified by a microwell binding assay. Briefly, diluted fractions were used to coat MaxiSorp plate wells and the HMGB1-binding capacity was measured by a recombinant HMGB1 binding assay as described [38].

The molecular weight of the HMGB1-binding platelet protein was determined using gel-filtration chromatography. Diethylaminoethanol-Sepharose –column eluted fractions that bound to HMGB1 were diluted five-fold with 5 mM octyl-glucoside in PBS and were applied to Mono Q column (GE Healthcare), washed with 0.15 M NaCl 10 mM phosphate, 5 mM octyl-glucoside, pH 7.5 and eluted with 0.15 – 2 M NaCl in the same buffer (0-100% gradient, 10 min, 1 ml/min, 1 ml fractions were collected) using ÄKTA Prime chromatography station (GE Healthcare). Absorbance at 280 nm and the relative conductivity of the eluate were monitored. The eluted protein-containing fractions were concentrated and fractionated using the ÄKTA Prime Superdex-75 gel filtration column chromatography. The mobile phase was 5 mM octyl-glucoside, 10 mM phosphate, 0.15 M NaCl, pH 7.5, at flow rate of 0.2 ml/min. 0.5 ml fractions were collected and their HMGB1-binding capacity was measured as described above.

In addition, HMGB1-binding platelet membrane anionic proteins were analysed using a HMGB1-affinity column chromatography and mass spectrometry. Anion exchange colum eluted fractions of platelet membrane proteins were diluted five-fold with 5 mM octyl-glucoside in PBS and applied to HMGB1-Sepahrose column. The column was washed with 5 mM octyl-glucoside in PBS and bound proteins were eluted with 5 mM octyl-gluco-side, 2 M NaCl in PBS and analysed in SDS-PAGE. The gel was stained with silver. The major high molecular weight band in the silver stained gel was analysed using mass spectrometry in the Proteomics Unit of Institute of Biotechnology (University of Helsinki, Finland). The eluted non-reduced protein fractions were also analysed in Western blot assay using anti-MMRN1 mouse monoclonal antibody ab56890 (Abcam, Cambridge, UK) as a primary antibody.

3.2. Proximity ligation assay

Proximity ligation assay was performed using Duolink II in situ PLA Protein Detection Kit (Olink Bioscience, Uppsala, Sweden). Human Umbilical Vein Endothelial Cells (HUVEC) –cells were obtained from PromoCell (Heidelberg, Germany) and human platelets were obtained from 3.8% citrate anticoagulated blood donated by healthy volunteers after an informed consent had been obtained according to the Declaration of Helsinki. The following antibodies were used: anti-MMRN1 mouse monoclonal antibody (Abcam), anti-HMG1 rabbit IgG (Pharmingen Becton Dickinson Co, Franklin Lakes, NJ, USA), anti-influenza hemagglutinin -probe (HA-probe) mouse monoclonal antibody (Santa Cruz Biotechnology, Santa Cruz, CA, USA) and rabbit anti-trimethyl-Histone H3 (Lys27) antibody (Millipore).

3.3. Platelet adhesion assay

Static adhesion of washed thrombin-activated mouse platelets to a protein coated microwell plate was measured as described [19, 41, 42].

3.4. Thrombin generation assay

Thrombin generation in solution was determined as described [43] using 80% phosphatidyl-choline / 20 % phosphatidylserine lipid vesicles prepared by sonication from 10 mg/ml of lipids (Avanti, Alabaster, Alabama, USA) in 3 mM $CaCl_2$-containing PBS –buffer. For the thrombin generation assay, vesicles were diluted and preincubated for 3 min at 37 °C in buffer B (137 mM NaCl, 10 mM Hepes, 5 mM glucose, 2.7 mM KCl, 2 mM $MgCl_2$, 0.05 % BSA, pH 7.4) with 3 mM $CaCl_2$. Coagulation factor mix with 5 nM Bovine Factor X/Xa (FX and FXa, Enzyme Research Labs Inc., Swansea, UK) and 10 nM Bovine Factor V/Va (FV and FVa, Enzyme Research Labs Inc.) in buffer B with $CaCl_2$ was added and preincubated for 1 min at 37°C prior to initiating the reaction with 10 μM bovine prothrombin (Enzyme Research Labs Inc.). Thrombin formation was assessed by subsampling 40 μl aliquots at selected intervals. Thrombin generation was terminated by the addition of 10 μl of stop buffer (120 mM NaCl, 50 mM Tris, 20 mM EDTA, 0.05 % BSA, pH 7.4). Thrombin activity was measured using a chromogenic substrate S-2238 (Chromogenix, Mölndal, Sweden) and the color reaction was stopped with acid and absorbance was measured at 405 nm. Recombinant HMGB1 was preincubated at a concentration of 5-50 μg/ml with the coagulation factor mix on ice. Final prothrombinase-activating conditions were 3 mM $CaCl_2$, 1.0 nM factor V/Va, 0.5 nM factor X/Xa and 1.0 μM prothrombin.

Thrombin generation on adherent platelets was determined as described [44]. Human platelets were isolated from healthy volunteers who denied having any medication for the previous 10 days. Blood was collected into 1/6 volumes of ACD-buffer (39 mM citric acid, 75 mM sodium citrate, 135 mM D-glucose, pH 4.5) and centrifuged at 200 g for 12 min. Platelet-rich plasma (PRP) was supplemented with 1/10 vol of ACD-buffer and 1/1000 vol of 100 ng/ml prosta-glandin E1 (Sigma-Aldhrich) for 15 min. Platelets were centrifuged for 15 min at 650 g, washed once and recentrifuged at 500 g for 15 min before their suspension into Tyrode's buffer (137 mM NaCl, 11.9 mM $NaHCO_3$, 2.7 mM KCl, 0.4 mM NaH_2PO_4, 1.1 $MgCl_2$ mM, 5.6 mM D-glucose, pH 7.4) with 0.35 % BSA. Platelet concentration was measured spectrofotometrically at 405 nm assuming A 0.025 to correspond to 1×10^6 cells/ml. Platelets were adhered at 100×10^6 cells/ml on collagen-coated (10 μg/ml, Kollagenreagens Horm, Hormon Chemie, Munich, Germany) for 1 hr at room temperature and washed three times before adding 280 ml of buffer B with 3 mM $CaCl_2$ for 30 min at room temperature, and followed by the thrombin generation assay as described above.

3.5. Release of HMGB1 from thrombin-activated washed mouse platelets

Blood was collected from anesthetised mice using cardiac puncture method and the blood was anticoagulated with 10 mM EDTA and 2 μg/ml of prostaglandin E1. Blood was centrifuged at 120 g for 5 minutes and PRP was collected. Two hundred microliters of CFT-buffer (135 mM

NaCl, 11.9 mM NaHCO$_3$, 0.4 mM Na$_2$HPO$_4$, 2.7 mM KCl, 5.6 mM Dextrose, pH 7.4) and 100 µl of Tyrode's buffer were added to rest of the blood and centrifuged again at 120 g for 5 minutes and PRP was collected. Again 200 µl of CFT-buffer and 100 µl of Tyrode's buffer was added to the rest of the blood. The blood was centrifuged at 120 g for 5 minutes and PRP was collected. Finally the PRP-fractions were pooled and 2 µg/ml of prostaglandin E1 was added. PRP was centrifuged at 2 000 g for 6 minutes and the cell pellet was washed twice with CFT - Tyrode's buffer containing prostaglandin E1. The washed platelet pellet was suspended to Tyrode's buffer containing 1 mM CaCl$_2$ and absorbance at 405 nm was measured to determine cell number as described above. Thrombin was added to cells (1 U/ml), cells were incubated at room temperature for 10 minutes and centrifuged at 2 000 g for 6 minutes. The platelet free supernatant was analysed with HMGB1 and mIL-6 ELISA Kits (IBL International GmbH, Hamburg, Germany).

4. Results

4.1. Binding of bacterial DNA to HMGB1-affinity chromatography column

The function of HMGB1-affinity column was validated by evaluating the column´s ability to bind bacterial DNA that is known to form proinflammatory complexes with HMGB1 [45]. Bacterial lysate was loaded to the column, the column was washed and the bound substances were eluted from the column with an increasing salt concentration. DNA in the eluted fractions was detected with a fluorescent DNA-dye. DNA bound strongly to the HMGB1-Sepharose column and did not bind at all to the control Sepharose-column lacking HMGB1 (Figure 1 A). We concluded that HMGB1-Sepharose affinity column in this study functioned in a similar way as the HMGB1-columns used in the previous studies [6, 46, 47].

4.2. Binding of histones H1 and H3 to HMGB1-affinity chromatography column

The function of HMGB1-affinity column was further validated by evaluating its ability to bind histones H1 and H3, since histones H1 and H3 have previously been shown to bind to HMGB1 in biochemical assays other than affinity chromatography [48, 49].

For affinity chromatography, nuclear extracts of C6 glioblastoma cells were loaded to a HMGB1-Sepharose column and the bound proteins were eluted with linear NaCl-gradient. Histones in elution fractions were analysed with ELISA. As a result histones H1 and H3 were detected among eluted proteins (Figure 1 B and C). In conclusion, it was shown that both histone H1 and H3 interact with HMGB1 and also affinity chromatography can be used to detect these interactions.

4.3. Identification of MMRN1-protein complex as a HMGB1-binding structure

Platelets are known to bind HMGB1, however, the cell surface receptors mediating the binding are poorly characterised. Here, we analysed the platelet membrane HMGB1-binding compo-nents using chromatographic, enzymatic, immunological and cell biological methods.

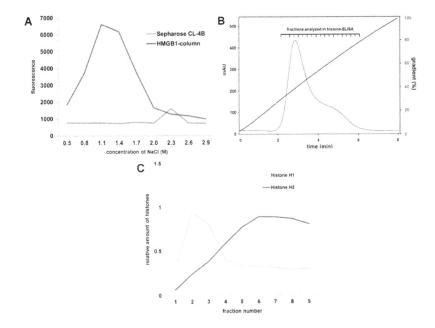

Figure 1. Binding of HMGB1 ligands to HMGB1-affinity chromatography column. A) Sepharose-coupled HMGB1 column was loaded with bacterial lysate and the bound substances were eluted with an increasing salt concentration.
Eluted material contained DNA that was detected with a fluorescent DNA-dye. B) HMGB1-affinity column was loaded
with a glioblastoma cell nuclear lysate and the bound proteins were eluted with linear NaCl-gradient. Dark grey line
indicates salt gradient, light gray line indicates absorbance at 280 nm. C) Protein containing fractions from the chromatography shown in figure 1 B were analysed with histone ELISA. Histones H1 and H3 were eluted from the column
with an increasing salt concentration. Histone H3 bound more strongly to the HMGB1-cloumn than did histone H1
(the fraction with the highest absorbance in each assay was determined as 1 and the relative values of other fractions
were calculated. N=3, mean values are shown).

Human platelet membrane proteins were isolated with anion exchange chromatography and
the isolated proteins were found to bind to recombinant HMGB1 in a microwell binding assay
(data not shown). HMGB1-binding proteins were further fractionated using gel filtration
chromatography and the binding of size-excluded proteins to recombinant HMGB1 was
determined. A high molecular weight protein fraction was found to bind to HMGB1 (Figure
2 A and B).

A similar high molecular weight protein fraction was found to contain endogenous HMGB1
from mouse platelets (Figure 5 A and B). When the anionic human platelet membrane protein
fraction was analysed in HMGB1-affinity chromatography, a high molecular weight protein
was found to be eluted from the column with high salt buffer (Figure 2 B). Mass spectrometric
analysis revealed that this protein had identical tryptic peptide sequences when compared to
Multimerin-1 (MMRN1) (Figure 2 C). Western blot analysis of the HMGB1-column eluted

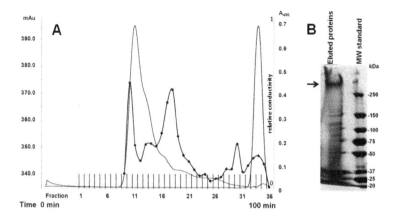

C

Match to: Q13201 Score: 39
Multimerin-1 OS=Homo sapiens GN=MMRN1 PE=1 SV=3

Start - End	Observed	Mr(expt)	Mr(calc)	ppm	Miss	Sequence
412 - 424	712.3538	1422.6930	1422.6838	6	0	K.TVSSLSEDLESTR.Q (Ions score 32)
796 - 804	541.2983	1080.5820	1080.5968	-14	0	R.YNFVLQVAK.T (Ions score 24)
951 - 961	646.3725	1290.7304	1290.7296	1	0	K.HSLPDIQLLQK.G (Ions score 25)
989 - 999	542.8240	1083.6335	1083.6288	4	0	R.SLPGSLANVVK.S (Ions score 22)

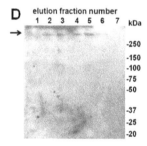

Figure 2. Isolation and identification of the HMGB1-binding component from platelets. A) Anionic platelet membrane proteins were fractionated with size-exclusion chromatography and the binding of recombinant HMGB1 to microwells coated with the chromatography fractions was analysed. The strongest HMGB1-binding was detected in the high molecular weight fraction. Black line in the chromatogram indicates absorbance at 490 nm in HMGB1-binding assay. Blue and red lines indicate absorbance at 280 nm (mAu) and relative conductivity, respectively. Fractions are indicated by short red vertical lines. B and C) Platelet membrane proteins were eluted from the HMGB1-affinity column and analysed by SDS-PAGE followed by mass spectrometry. The high molecular weight protein band indicated by the arrow was identified as Multimerin-1. D) A Western blotting analysis of platelet membrane proteins eluted with high salt from HMGB1-affinity column. The arrow points to the high molecular weight band detected with an anti-Multimerin-1 antibody.

anionic platelet membrane proteins with anti-MMRN1 monoclonal antibody confirmed the finding that MMRN1 bound to HMGB1-column (Figure 2 D).

Since both HMGB1 and MMRN1 are present on the surface of activated platelets and cultured endothelial cells, we studied the association of HMGB1 and MMRN1 on the cell surface using a proximity ligation assay. With the proximity ligation assay, it could be demonstrated that HMGB1 and MMRN1 are in close proximity of each other on both the activated platelet and the cultured endothelial cell surfaces (Figure 3).

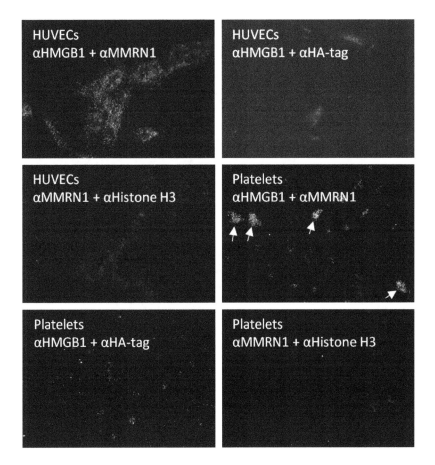

Figure 3. Endogenous HMGB1 associates with MMRN1. Proxomity ligation assay with anti-HMGB1 and anti-MMRN1 antibodies revealed a close association of HMGB1 and Multimerin-1 on both the endothelial cell and the activated platelet surfaces (arrows indicate positively stained platelets). Negative control stained cells with anti-trimethyl Histone H3 and anti-MMRN1 antibodies or with anti-HA-probe and anti-HMGB1 antibodies did not yield significant signals in the proxomity ligation assay.

MMRN1 has been shown to enhance platelet adhesion to collagen. Next, we tested whether MMRN1 mediates static adhesion of activated platelets to HMGB1. We used both platelets derived from control C57BL/6JOlaHsd mice deficient in Multimerin-1 and α-synuclein genes and platelets derived from C57BL/N6 mice expressing Multimerin-1 and α-synuclein [50]. We could not observe any significant differences in platelet adhesion to HMGB1 between the two different mouse strains (data not shown).

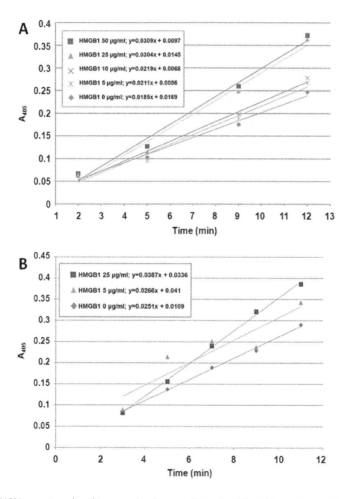

Figure 4. HMGB1 potentiates thrombin generation. Factor Va/Xa mediated thrombin generation was potentiated by recombinant HMGB1 in a similar way on both phospholipid vesicles (A) and the activated platelet surfaces (B). Potentiation occurred in both cases at ≥25 µg/ml concentration of recombinant HMGB1. 0 = control with PBS as a vehicle. A representative graph of three repetitions is shown.

Both MMRN1 and HMGB1 bind to phosphatidylserine [19, 38, 51]. Previously, it has been shown that HMGB1 enhances the effects of thrombin *in vivo* [52]. In contrast, MMRN1 has been shown to inhibit thrombin generation in plasma *in vitro* [53]. Therefore, we tested whether HMGB1 affects FXa- and FVa-catalysed thrombin generation on phosphatidylcholine/phosphatidylserine vesicles and on activated human platelets. Addition of exogenous HMGB1 enhanced thrombin generation on both the vesicle and the activated platelet surfaces (Figure 4 A and B). The minimal concentration of HMGB1 needed for enhancement was 25 μg/ml on both surfaces suggesting that the potentiation of thrombin generation by HMGB1 is mediated by similar mechanism in both systems, ie. the potentiation requires a lipid surface, but it is independent of other structures, including MMRN1 present on the activated platelet surface.

We further evaluated the possible role of MMRN1 on the biology of HMGB1. Gel filtration assays of platelet protein lysates from mouse platelets lacking MMRN1 and α-synuclein genes and from mouse platelets having MMRN1 and α-synuclein genes revealed that the sendogenous HMGB1 is mainly complexed with high molecular weight structures indicating a MMRN1-independent high molecular weight HMGB1-complex formation (Figure 5 A and B). Finally, the release of HMGB1 from thrombin-activated washed mouse platelets was measured with an ELISA. The concentration of HMGB1 released from 1 U/ml of thrombin-activated washed C57BL/N6 mouse platelets was 5.7 ± 0.8 ng/ml (n=4). We observed a significant difference in the concentrations of the 1 U/ml of thrombin-activated platelet released HMGB1 between the mice lacking MMRN1 and α-synuclein and the mice having MMRN1 and α-synuclein (Figure 5 C). Platelets from the mice lacking MMRN1 and α-synuclein genes released more HMGB1 than platelets from the mice having MMRN1 and α-synuclein genes. Platelets from mice lacking MMRN1 and α-synuclein genes released more HMGB1 than platelets from mice having MMRN1 and α-synuclein genes (Figure 5 C). α-Synyclein regulates vesicle release in many cell types. However, measurement of IL-6 release did not show any difference between the two mouse strains, which is corroborated by the finding that the activation of both types of mouse platelets with 1 U/ml of thrombin induced similar release of P-selectin [54]. This suggests that the observed difference in HMGB1 release may not be due to a differential release response to a high concentration of thrombin. HMGB1 has been described to bind to the filamentous pathological form of α-Synyclein but not to monomers [55]. However, since α-synuclein occurs in a monomeric form in the platelets [56], the lack of α-synuclein hardly affects the release of HMGB1.

5. Conclusions

In this study, we isolated a MMRN1-complex with HMGB1-affinity chromatography from platelets. MMRN1 and HMGB1 were shown to exist at a close proximity of each other on both the endothelial cell and the activated platelet surface, and the release of HMGB1 from activated platelets from mice lacking MMRN1 and α-synuclein genes was increased. Further, HMGB1 potentiated FVa/FXa catalysed thrombin generation on both artificial anionic vesicles and the activated platelet surface.

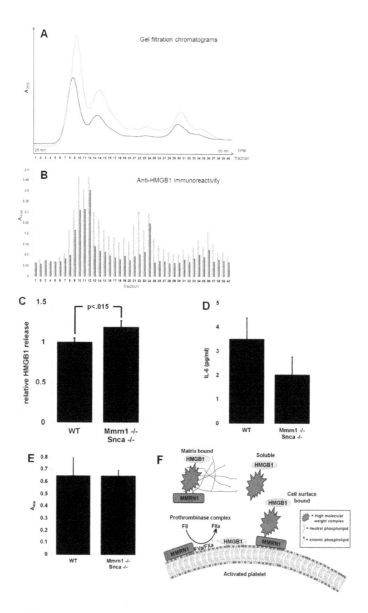

Figure 5. Size-exclusion chromatography and the secretion analyses of HMGB1 using washed platelets derived either from Mmrn1/Snca negative C57BL/6JOlaHsd mouse or C57BL/N6 mouse expressing MMRN1/SNCA. A) Gel filtration of mouse platelet anionic proteins. Light grey line indicates platelet proteins from C57BL/6JOlaHsd mouse and dark grey line indicates platelet proteins from C57BL/N6 mouse. B) Binding of anti-HMGB1 antibody to the microwells coated with the gel filtration eluted fractions. Light grey bars indicate the fractions from C57BL/6JOlaHsd mouse pla-

telets and dark grey bars indicate the fractions from C57BL/N6 mouse platelets. C) Higher concentration of HMGB1 is detected in platelet released material derived from 1 U/ml of thrombin-activated mouse platelets lacking MMRN1/ SNCA when compared to released material derived from 1 U/ml of thrombin-activated mouse platelets expressing Mmrn1/Snca (n=6 C57BL/6JOlaHsd mice, n=8 C57BL/N6 mice; mean HMGB1 amount in C57BL/N6 derived samples was determined as 1 and the relative values were calculated). HMGB1 levels were at the linear range of ELISA detection. D) In contrast to HMGB1-release, there was no difference in interleukin-6 release from either type of platelets. E) Absorbance at 405 nm was equal in washed platelet fractions derived from either strain indicating the existence of equal number of cells (n=4). F) A hypothetical model of association of HMGB1 to the activated platelet surface and matrix. Activated platelets express phosphatidylserine on their surface. The prothrombinase complex consisting of co-agulation factors FVa and FXa, MMRN1 and HMGB1 can all bind to phosphatidylserine. Phosphatidylserine surfaces catalyse the formation of thrombin (FIIa) from prothrombin (FII). HMGB1 can potentiate thrombin generation. Activated platelets release HMGB1 that can be found in the media either as a soluble molecule or bound to microvesicles. It can also bind to the extracellular matrix and the cell surfaces. The release of HMGB1 from activated platelets was found to be increased in the platelets derived from Mmrn1/Snca negative mice. Possible explanations for this phenomenon are that MMRN1 complex anchors HMGB1 to the cell surface or it interferes with HMGB1-ELISA detection like some other HMGB1 binding components (60).

MMRN1 is a variably-sized homopolymer belonging to the disulphide–linked multimeric proteins of the elastin microfibril interface located protein (EMILIN) family. Megakaryocytes, platelets and endothelial cells carry MMRN1 in their secretory granules, α-granules or Weibel-Palade bodies, respectively. MMRN1 size can range from trimers to large disulphide linked polymers which can exceed millions of Daltons [57, 58]. The prepro-MMRN1 molecule contains 1228 amino acids with a 19 amino acid signal peptide. MMRN1becomes only liberated upon cell activation and has not been detected as a free plasma molecule. MMRN1 has functions in both platelet adhesion and procoagulant activity. As an interesting link to immunity, MMRN1 has also been shown to mediate neutrophil binding [59].

The adhesive function of MMRN1 is manifested in its several molecular features (for a review see reference [60]). MMRN1 can assemble into fibrils and become associated to extracellular matrix proteins such as collagens type I and type III [62]. MMRN1 contains the RGD-sequence by which it can interact with e.g. integrins $\alpha_{IIb}\beta_3$ or $\alpha_v\beta_3$, but integrin-independent binding can also occur via phosphatidyl serine [61]. MMRN1 has been shown to support platelet adhesion even at high shear rates thus resembling the function of von Willebrand factor [62]. Additionally, MMRN1 has been shown to increase von Willebrand factor -dependent platelet adhesion to collagen.

In addition to platelet adhesion, MMRN1 participates in procoagulant activity. Like FVa, MMRN1 binds to phospholipids [51]. MMRN1-lipid binding was enhanced by increasing phosphatidylserine content of phosphatidylserine:phosphatidylethanolamine membranes, and by increasing phosphatidylethanolamine and cholesterol content of low phosphatidylserine membranes [60]. Additionally, MMRN1 binds both inactive and active factor V with high affinity (Kd 2 and 7 nM), but surprisingly, its role in thrombin generation has been suggested to be inhibitory [53]. Exogenous MMRN1 has been shown to delay and reduce thrombin generation by plasma and platelets. In this capacity, MMRN1 could act as controller of unlimited thrombin generation upon vascular injury. However, the effect of MMRN1 on FVa function *in vivo* has not been explored.

MMRN1 gene and α-synuclein gene knockout mice were used to study Ferric chloride – induced thrombus formation with intravital microscopy. In these mice, platelet adhesion and thrombus formation were impaired and the deficit could be corrected with exogenous MMRN1 [54].

No specific MMRN1 deficiency in humans or animals has been reported. MMRN1 deficiency is not lethal since a deletion combining the major part of MMRN1 gene and a-synuclein in mice renders them viable and without an obvious phenotype [50]. In humans, MMRN1 gene has been linked to Parkinson's disease and neurodevelopmental disorders [63, 64]. In addition, a genetic multiplication of SNCA and MMRN1 locus in humans can lead to parkinsonism [65]. Whereas α-synuclein gene is highly expressed in the brain, the expression of MMRN1 gene can be detected in neural stem cells [65, 66].

Our results show that endogenous HMGB1 associates to a high molecular weight MMRN1-complex in human platelets. However, HMGB1 also associates to a high molecular weight complex in mouse platelets lacking MMRN1, suggesting that MMRN1 itself does not mediate the HMGB1-interactions within the complex. MMRN1 can instead mediate interactions of the complex with the activated platelet surface leading to a decreased amount of platelet released soluble HMGB1 (Figure 5). The mechanism of the HMGB1 induced potentiation of thrombin formation observed in this study remains unknown. Both the prothrombinase complex and HMGB1 bind to phosphatidylserine. However, whether HMGB1 is in direct contact with coagulation factors requires further investigation.

Acknowledgements

Ari Rouhiainen was supported by the Orion-Farmos Research Foundation. Pia Siljander was supported by the Magnus Ehrnrooth Foundation. Heikki Rauvala was supported by the Academy of Finland and the Sigrid Jusélius Foundation.

Author details

Ari Rouhiainen[1], Helena Tukiainen[2], Pia Siljander[2] and Heikki Rauvala[1]

1 Neuroscience Center University of Helsinki, Finland

2 Deparment of Biosciences, University of Helsinki, Finland

References

[1] Rauvala H, Pihlaskari R. Isolation and some characteristics of an adhesive factor of brain that enhances neurite outgrowth in central neurons. *Journal of Biological Chemistry*. (1987). 262(34), 16625-16635.

[2] Merenmies J, Pihlaskari R, Laitinen J, Wartiovaara J, Rauvala H. 30-kDa heparin-binding protein of brain (amphoterin) involved in neurite outgrowth. Amino acid sequence and localization in the filopodia of the advancing plasma membrane. *Journal of Biological Chemistry*. (1991). 266(25), 16722-16729.

[3] Bianchi ME, Beltrame M, Paonessa G. Specific recognition of cruciform DNA by nuclear protein HMG1. *Science*. (1989). 243(4894 Pt 1), 1056-1059.

[4] Bustin M. Revised nomenclature for high mobility group (HMG) chromosomal proteins. *Trends in Biochemical Sciences*. (2001). 26(3), 152-153.

[5] Hori O, Brett J, Slattery T, Cao R, Zhang J, Chen JX, Nagashima M, Lundh ER, Vijay S, Nitecki D, Morser J, Stern D, Schmidt AM. The receptor for advanced glycation end products (RAGE) is a cellular binding site for amphoterin. Mediation of neurite outgrowth and co-expression of rage and amphoterin in the developing nervous system. *Journal of Biological Chemistry*. (1995). 270(43), 25752-25761.

[6] Isackson PJ, Fishback JL, Bidney DL, Reeck GR. Preferential affinity of high molecular weight high mobility group non-histone chromatin proteins for single-stranded DNA. *Journal of Biological Chemistry*. (1979). 254(13), 5569-5572.

[7] Chou DK, Evans JE, Jungalwala FB. Identity of nuclear high-mobility-group protein, HMG-1, and sulfoglucuronyl carbohydrate-binding protein, SBP-1, in brain. *Journal of Neurochemistry*. (2001). 77(1), 120-131.

[8] Srikrishna G, Huttunen HJ, Johansson L, Weigle B, Yamaguchi Y, Rauvala H, Freeze HH. N-Glycans on the receptor for advanced glycation end products influence amphoterin binding and neurite outgrowth. *Journal of Neurochemistry*. (2002). 80(6), 998-1008.

[9] Rauvala H, Rouhiainen A. Physiological and pathophysiological outcomes of the interactions of HMGB1 with cell surface receptors. *Biochimica et Biophysica Acta*. (2010). 1799(1-2), 164-170.

[10] Thomas JO, Stott K. H1 and HMGB1: modulators of chromatin structure. *Biochemical Society Transactions*. (2012). 40(2), 341-346.

[11] Raulo E, Chernousov MA, Carey DJ, Nolo R, Rauvala H. Isolation of a neuronal cell surface receptor of heparin binding growth-associated molecule (HB-GAM). Identification as N-syndecan (syndecan-3). *Journal of Biological Chemistry*. (1994). 269(17), 12999-3004.

[12] Zappavigna V, Falciola L, Helmer-Citterich M, Mavilio F, Bianchi ME. HMG1 interacts with HOX proteins and enhances their DNA binding and transcriptional activation. *EMBO Journal*. (1996). 15(18), 4981-4991.

[13] Huttunen HJ, Fages C, Kuja-Panula J, Ridley AJ, Rauvala H. Receptor for advanced glycation end products-binding COOH-terminal motif of amphoterin inhibits invasive migration and metastasis. *Cancer Research*. (2002). 62(16), 4805-4811.

[14] Wang H, Bloom O, Zhang M, Vishnubhakat JM, Ombrellino M, Che J, Frazier A, Yang H, Ivanova S, Borovikova L, Manogue KR, Faist E, Abraham E, Andersson J, Andersson U, Molina PE, Abumrad NN, Sama A, Tracey KJ. HMG-1 as a late mediator of endotoxin lethality in mice. *Science*. (1999). 285(5425), 248-251.

[15] Lotze MT, Tracey KJ. High-mobility group box 1 protein (HMGB1): nuclear weapon in the immune arsenal. *Nature Reviews Immunology*. (2005). 5(4), 331-342.

[16] Sims GP, Rowe DC, Rietdijk ST, Herbst R, Coyle AJ. HMGB1 and RAGE in inflammation and cancer. *Annual Review of Immunology*. (2010). 28, 367-388.

[17] Rouhiainen A, Kuja-Panula J, Tumova S, Rauvala H. RAGE-Mediated Cell Signaling. *Methods in Molecular Biology*. (2013). 963, 239-263. doi:10.1007/978-1-62703-230-8_15.

[18] Rouhiainen A, Kuja-Panula J, Wilkman E, Pakkanen J, Stenfors J, Tuominen RK, Lepäntalo M, Carpén O, Parkkinen J, Rauvala H. Regulation of monocyte migration by amphoterin (HMGB1). *Blood*. (2004). 104(4), 1174-1182.

[19] Rouhiainen A, Imai S, Rauvala H, Parkkinen J. Occurrence of amphoterin (HMG1) as an endogenous protein of human platelets that is exported to the cell surface upon platelet activation. *Thrombosis and Haemostasis*. (2000). 84(6), 1087-1094.

[20] Mullins GE, Sunden-Cullberg J, Johansson AS, Rouhiainen A, Erlandsson-Harris H, Yang H, Tracey KJ, Rauvala H, Palmblad J, Andersson J, Treutiger CJ. Activation of human umbilical vein endothelial cells leads to relocation and release of high-mobility group box chromosomal protein 1. *Scandinavian Journal of Immunology*. (2004). 60(6), 566-573.

[21] Hofner P, Seprényi G, Miczák A, Buzás K, Gyulai Z, Medzihradszky KF, Rouhiainen A, Rauvala H, Mándi Y. High mobility group box 1 protein induction by Mycobacterium bovis BCG. *Mediators of Inflammation*. (2007). doi:10. 1155/2007/53805.

[22] Maugeri N, Franchini S, Campana L, Baldini M, Ramirez GA, Sabbadini MG, Rovere-Querini P, Manfredi AA. Circulating platelets as a source of the damage-associated molecular pattern HMGB1 in patients with systemic sclerosis. *Autoimmunity*. (2012). 45(8), 584-587.

[23] Andersson U, Tracey KJ. HMGB1 is a therapeutictarget for sterileinflammation and infection. *Annual Review of Immunology*. (2011). 29, 139-162.

[24] Ito T, Kawahara K, Okamoto K, Yamada S, Yasuda M, Imaizumi H, Nawa Y, Meng X, Shrestha B, Hashiguchi T, Maruyama I. Proteolytic cleavage of high mobility group box 1 protein by thrombin-thrombomodulin complexes. *Arteriosclerosis, Thrombosis, and Vascular Biology*. (2008). 28(10), 1825-1830.

[25] Antoine DJ, Jenkins RE, Dear JW, Williams DP, McGill MR, Sharpe MR, Craig DG, Simpson KJ, Jaeschke H, Park BK. Molecular forms of HMGB1 and keratin-18 as mechanistic biomarkers for mode of cell death and prognosis during clinical acetaminophen hepatotoxicity. *Journal of Hepatology*. (20120., 56(5), 1070-1079.

[26] Urbonaviciute V, Meister S, Fürnrohr BG, Frey B, Gückel E, Schett G, Herrmann M, Voll RE. Oxidation of the alarmin high-mobility group box 1 protein (HMGB1) during apoptosis. *Autoimmunity*. (2009). 42(4), 305-307.

[27] Semple JW, Italiano JE Jr, Freedman J. Platelets and the immune continuum. *Nature Reviews Immunology*. (2011). 11(4), 264-274.

[28] Echtler K, Stark K, Lorenz M, Kerstan S, Walch A, Jennen L, Rudelius M, Seidl S, Kremmer E, Emambokus NR, von Bruehl ML, Frampton J, Isermann B, Genzel-Boroviczény O, Schreiber C, Mehilli J, Kastrati A, Schwaiger M, Shivdasani RA, Massberg S. Platelets contribute to postnatal occlusion of the ductusarteriosus. *Nature Medicine*. (2010). 16(1), 75-82.

[29] Horstman LL, Jy W, Ahn YS, Zivadinov R, Maghzi AH, Etemadifar M, Steven Alexander J, Minagar A. Role of platelets in neuroinflammation: a wide-angle perspective. *Journal of Neuroinflammation*. (2010). doi:10. 1186/1742-2094-7-10.

[30] Manfredi AA, Rovere-Querini P, Maugeri N. Dangerous connections: neutrophils and the phagocytic clearance of activated platelets. *Current Opinion in Hematology*. (2010). 17(1), 3-8.

[31] Piersma SR, Broxterman HJ, Kapci M, de Haas RR, Hoekman K, Verheul HM, Jiménez CR. Proteomics of the TRAP-induced platelet releasate. *Journal of Proteomics*. (2009). 72(1), 91-109.

[32] Zhu W, Li W, Silverstein RL. Advanced glycation end products induce a prothrombotic phenotype in mice via interaction with plateletCD36. *Blood*. (2012). 119(25), 6136-6144.

[33] Okayama M, Oguri K, Fujiwara Y, Nakanishi H, Yonekura H, Kondo T, Ui N. Purification and characterization of human platelet proteoglycan. *Biochemical Journal*. (1986). 233(1), 73-81.

[34] AhrensI,Agrotis A,Topcic D,Bassler N, Chen Y,Bobik A, Bode C, Peter K. The Pro-Atherogenic DNA-Binding Cytokine HMGB1 Binds to Activated Platelets via the Receptor for Advanced Glycation End Products (RAGE) and Toll-Like Receptor 2 (TLR 2). *Circulation*. (2011). 124, A13245.

[35] Hagiwara S, Iwasaka H, Hidaka S, Hishiyama S, Noguchi T. Danaparoid sodium inhibits systemic inflammation and prevents endotoxin-induced acute lung injury in rats. *Critical Care*. (2008). 12(2), R43.

[36] Hagiwara S, Iwasaka H, Hasegawa A, Oyama M, Imatomi R, Uchida T, Noguchi T. Adenosine diphosphate receptor antagonist clopidogrelsulfate attenuates LPS-induced systemic inflammation in a rat model. *Shock*. (2011). 35(3), 289-292.

[37] Nomura S, Fujita S, Ozasa R, Nakanishi T, Miyaji M, Mori S, Ito T, Ishii K. The correlation between platelet activation markers and HMGB1 in patients with disseminat-

ed intravascular coagulation and hematologic malignancy. *Platelets.* (2011). 22(5), 396-397.

[38] Rouhiainen A, Tumova S, Valmu L, Kalkkinen N, Rauvala H. Analysis of proinflammatory activity of highly purified eukaryotic recombinant HMGB1 (amphoterin). *Journal Leukocyte Biology.* (2007). 81(1), 49-58.

[39] Benda P, Lightbody J, Sato G, Levine L, Sweet W. Differentiated rat glial cell strain in tissue culture. *Science.* (1968). 161(3839), 370-371.

[40] Díaz-Jullien C, Pérez-Estévez A, Covelo G, Freire M. Prothymosin alpha binds histones in vitro and shows activity in nucleosome assembly assay. *Biochimica et Biophysica Acta.* (1996). 1296(2), 219-227.

[41] Bellavite P, Andrioli G, Guzzo P, Arigliano P, Chirumbolo S, Manzato F, Santonastaso C. A colorimetric method for the measurement of platelet adhesion in microtiter plates. *Analytical Biochemistry.* (1994). 216(2), 444-450.

[42] Ilveskero S, Siljander P, Lassila R. Procoagulant activity on platelets adhered to collagen or plasma clot. *Arteriosclerosis, Thrombosis, and Vascular Biology.* (2001). 21(4), 628-635.

[43] Heemskerk JW, Comfurius P, Feijge MA, Bevers EM. Measurement of the platelet procoagulant response. *Methods in Molecular Biology.* (2004). 272, 135-144.

[44] Siljander P, Carpen O, Lassila R. Platelet-derived microparticles associate with fibrin during thrombosis. *Blood.* (1996). 87(11), 4651-63.

[45] Tian J, Avalos AM, Mao SY, Chen B, Senthil K, Wu H, Parroche P, Drabic S, Golenbock D, Sirois C, Hua J, An LL, Audoly L, La Rosa G, Bierhaus A, Naworth P, Marshak-Rothstein A, Crow MK, Fitzgerald KA, Latz E, Kiener PA, Coyle AJ. Tolllikereceptor 9-dependent activation by DNA-containing immune complexes is mediated by HMGB1 and RAGE. *Nature Immunology.* (2007). 8(5), 487-496.

[46] Watson M, Stott K, Thomas JO. Mapping intramolecular interactions between domains in HMGB1 using a tail-truncation approach. *Journal of Molecular Biology.* (2007). 374(5), 1286-1297.

[47] Ivanov S, Dragoi AM, Wang X, Dallacosta C, Louten J, Musco G, Sitia G, Yap GS, Wan Y, Biron CA, Bianchi ME, Wang H, Chu WM. A novel role for HMGB1 in TLR9-mediated inflammatory responses to CpG-DNA. *Blood.* (2007). 110(6), 1970-1981.

[48] Cato L, Stott K, Watson M, Thomas JO. The interaction of HMGB1 and linker histones occurs through their acidic and basictails. *Journal of Molecular Biology.* (2008). 384(5), 1262-1272.

[49] Ueda T, Chou H, Kawase T, Shirakawa H, Yoshida M. Acidic C-tail of HMGB1 is required for its target binding to nucleosome linker DNA and transcription stimulation. *Biochemistry.* (2004). 43(30), 9901-9908.

[50] Specht CG, Schoepfer R. Deletion of multimerin-1 in alpha-synuclein-deficient mice. *Genomics*. (2004). 83(6), 1176-1178.

[51] Jeimy SB, Woram RA, Fuller N, Quinn-Allen MA, Nicolaes GA, Dahlbäck B, Kane WH, Hayward CP. Identification of the MMRN1 binding region within the C2 domain of human factor V. *Journal of Biological Chemistry*. (2004). 279(49), 51466-51471.

[52] Ito T, Kawahara K, Nakamura T, Yamada S, Nakamura T, Abeyama K, Hashiguchi T, Maruyama I. High-mobility group box 1 protein promotes development of microvascular thrombosis in rats. *Journal of Thrombosis and Haemostasis*. (2007). 5(1), 109-116.

[53] Jeimy SB, Fuller N, Tasneem S, Segers K, Stafford AR, Weitz JI, Camire RM, Nicolaes GA, Hayward CP. Multimerin 1 binds factor V and activated factor V with high affinity and inhibits thrombin generation. *Thrombosis and Haemostasis*. (2008). 100(6), 1058-1067.

[54] Reheman A, Tasneem S, Ni H, Hayward CP. Mice with deleted multimerin 1 and alpha-synuclein genes have impaired platelet adhesion and impaired thrombus formation that is corrected by multimerin 1. *Thrombosis Research*. (2010). 125(5), e177-183.

[55] Lindersson EK, Højrup P, Gai WP, Locker D, Martin D, Jensen PH. alpha-Synuclein filaments bind the transcriptional regulator HMGB-1. *Neuroreport*. (2004). 15(18), 2735-2739.

[56] Li QX, Campbell BC, McLean CA, Thyagarajan D, Gai WP, Kapsa RM, Beyreuther K, Masters CL, Culvenor JG. Platelet alpha- and gamma-synucleins in Parkinson's disease and normal control subjects. *Journal of Alzheimer's Disease*. (2002). 4(4), 309-315.

[57] Hayward CP, Cramer EM, Song Z, Zheng S, Fung R, Massé JM, Stead RH, Podor TJ. Studies of multimerin in human endothelial cells. *Blood*. (1998). 91(4), 1304-1317.

[58] Hayward CP, Bainton DF, Smith JW, Horsewood P, Stead RH, Podor TJ, Warkentin TE, Kelton JG. Multimerin is found in the alpha-granules of resting platelets and is synthesized by a megakaryocytic cell line. *Journal of Clinical Investigation*. (1993). 91(6), 2630-2639.

[59] Urbonaviciute V, Fürnrohr BG, Weber C, Haslbeck M, Wilhelm S, Herrmann M, Voll RE. Factors masking HMGB1 in human serum and plasma. *J Leukoc Biol*. (2007). 81(1), 67-74.

[60] Jeimy SB, Tasneem S, Cramer EM, Hayward CP. Multimerin 1. *Platelets*. (2008). 19(2), 83-95.

[61] Adam F, Zheng S, Joshi N, Kelton DS, Sandhu A, Suehiro Y, Jeimy SB, Santos AV, Massé JM, Kelton JG, Cramer EM, Hayward CP. Analyses of cellular multimerin 1 receptors: in vitro evidence of binding mediated by alphaIIbbeta3 and alphavbeta3. *Thrombosis and Haemostasis*. (2005). 94(5), 1004-1011.

[62] Tasneem S, Adam F, Minullina I, Pawlikowska M, Hui SK, Zheng S, Miller JL, Hayward CP. Platelet adhesion to multimerin 1 in vitro: influences of platelet membrane

receptors, von Willebrand factor and shear. *Journal of Thrombosis and Haemostasis.* (2009). 7(4), 685-692.

[63] Iossifov I, Zheng T, Baron M, Gilliam TC, Rzhetsky A. Genetic-linkage mapping of complex hereditary disorders to a whole-genome molecular-interaction network. *Genome Research.* (2008). 18(7), 1150-1162.

[64] Zou F, Chai HS, Younkin CS, Allen M, Crook J, Pankratz VS, Carrasquillo MM, Rowley CN, Nair AA, Middha S, Maharjan S, Nguyen T, Ma L, Malphrus KG, Palusak R, Lincoln S, Bisceglio G, Georgescu C, Kouri N, Kolbert CP, Jen J, Haines JL, Mayeux R, Pericak-Vance MA, Farrer LA, Schellenberg GD; Alzheimer's Disease Genetics Consortium, Petersen RC, Graff-Radford NR, Dickson DW, Younkin SG, Ertekin-Taner N. Brain expression genome-wide association study (eGWAS) identifies human disease-associated variants. *PLoS Genetics.* (2012). 8(6), e1002707.

[65] Fuchs J, Nilsson C, Kachergus J, Munz M, Larsson EM, Schüle B, Langston JW, Middleton FA, Ross OA, Hulihan M, Gasser T, Farrer MJ. Phenotypic variation in a large Swedish pedigree due to SNCA duplication and triplication. *Neurology.* (2007). 68(12), 916-922.

[66] Koch P, Opitz T, Steinbeck JA, Ladewig J, Brüstle O. A rosette-type, self-renewing human ES cell-derived neural stem cell with potential for in vitro instruction and synaptic integration. *Proceedings of the National Academy of Sciences.* (2009). 106(9), 3225-3230.

Protein Engineering of
Enzymes Involved in Bioplastic Metabolism

Tomohiro Hiraishi and Seiichi Taguchi

Additional information is available at the end of the chapter

1. Introduction

1.1. Environmental problems caused by petroleum-based plastics

The last half century has witnessed the development of synthetic plastics from petroleum resources, and more than 300 million tons of synthetic plastics are annually produced at present. The recently increased consumption of petroleum resources has led to environmental problems such as a depletion of the resources as well as a global warming due to a marked increase of atmospheric CO_2 level. In addition to these problems, wasted plastics used in short-term applications may cause the environmental damage to a wide variety of wild animals including terrestrial, aquatic animals and birds. Furthermore, it has been suggested that even the wasted plastics in the form of small particles potentially induce the alteration of pelagic ecosystems [1]. Therefore, the development of environmentally sound alternatives, such as bioplastics, to conventional petroleum-based plastics is urgently desired to sustain the environment [2-4].

1.2. Bioplastics

Bioplastics include biodegradable and bio-based plastics (Figure 1) [5, 6]. The former are produced from renewable or petroleum resources via biological or chemical processes, and degraded by enzymes and microorganisms in natural environment. The latter are synthesized from renewable resources via biological or chemical processes, and some of them show non-biodegradability although bio-based plastics are generally biodegradable. Poly(ε-caprolactone) (PCL), poly(ethylene succinate) (PES) and poly(butylene succinate) (PBS) are synthesized from petroleum resources via chemical processes, but they show an excellent biodegradability. Currently, cost-effective processes for the production of succinic acid and 1,4-butanediol, raw materials of PBS, from biomass resources are being developed.

Meanwhile, poly(ethylene) (PE) and poly(propylene) (PP) are chemically synthesized from their monomers derived from biological sources, but they are not biodegradable. Poly(hydroxyalkanoate)s (PHAs) and poly(lactide) (PLA) show an excellent biodegradability, and are produced from renewable resources via biological and chemical processes, respectively. Thus, the bio-based bioplastics having biodegradability, such as PHAs and PLA, are the most favorable bioplastics to avoid the above-mentioned problems associated with the use of petrochemical-based synthetic plastics.

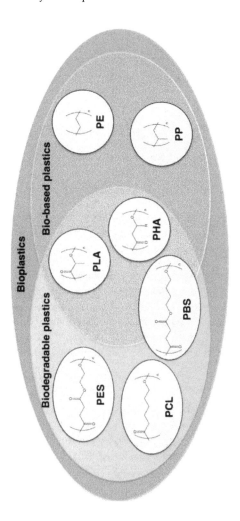

T. Hiraishi and S. Taguchi

Figure 1. Bioplastics comprised of biodegradable and bio-based plastics.

1.2.1. PHAs

PHAs are the only bioplastics completely synthesized from renewable resources by a wide variety of microorganisms in soil, active sludge, marine and extreme environments [7, 8]. In the cells, PHAs form amorphous granules and is degraded by intracellular PHA depolymerases (i-PHA depolymerases) produced by the PHA-accumulating bacterium itself. In contrast, after PHAs are extracted from the cells, PHAs are converted to semicrystalline form and is degraded by extracellular PHA depolymerases (e-PHA depolymerases) secreted from microorganisms in natural environments, such as soil, active sludge, fresh water, and seawater [9, 10].

Many bacteria can synthesize various types of PHAs containing 3-, 4-, and 5-hydroxyalkanoate units, and over 150 different hydroxyalkanoates other than 3-hydroxybutyrate have been reported as constitutive units of PHAs [11]. PHAs consisting of short-chain hydroxyalkanoates (SCL-HAs; 3–5 carbon atoms) or medium-chain hydroxyalkanoates (MCL-HAs; 6–14 carbon atoms) have been detected. The former are thermoplastic in nature, whereas, the latter are elastomeric in nature. The physical and mechanical properties of PHAs can be regulated by varying monomer composition in order to gain properties comparable to petrochemical-based thermoplastics that have been used for various applications in industry, medicine, pharmacy, agriculture, and electronics [12]. Accordingly, PHAs have attracted industrial interest as bio-based, biodegradable, biocompatible, and versatile thermoplastics [13, 14].

1.2.2. PLA

PLA is representative bio-based plastics with good processability and transparency that are used in packaging, containers, stationary, etc. [15]. In addition, medical and agricultural uses of the material have been investigated because of their biocompatibility and biodegradability [16]. PLAs are produced from renewable biomass through a chemo-bioprocess consisting of fermentative production of lactic acid (LA) and chemical polymerization. LA is spontaneously polymerized by refluxing, but the molecular mass of yielded polymer tends to be low [17]. There are several methods for synthesizing high-molecular-mass PLAs: condensation, chain elongation, and ring-opening polymerization of cyclic lactides [15]. Currently, the major industrial method to produce PLAs is ring-opening polymerization which is catalyzed by heavy metal catalysts, typically tin [18, 19]. However, the trace residues of the heavy metal catalyst are unfavorable for certain applications, in particular, medical and food applications. Thus, replacement of the heavy metal catalyst with a safe and environmentally acceptable alternative is an important issue. For this purpose, enzymes are attractive targets because they are natural non-harmful catalysts that can drive the reactions under mild conditions. In addition, highly specific enzymatic reactions may be capable of synthesizing polymers with fine structure from crude materials, which would reduce the cost of preparing the starting substances. This could be an advantage over chemical polymerization of LA or lactides, since the chemoprocess requires extremely pure monomers (contamination of carbonic acids is known to inhibit polymerization), along with anhydrous and high temperature conditions to proceed.

In such a situation, LA-polymerizing enzyme (LPE) functioning in replacement of metal catalysts should enable the biosynthesis of PLA, even though it is enormously challenging both in terms of research and industrial implementation. The best solution could be the development of a PLA-producing microorganism introduced with LPE gene, but this has not been reported so far. In 2008, Taguchi et al. nonetheless successfully obtained encouraging results by developing a recombinant *Escherichia coli* strain allowing the synthesis of LA-based polyesters by introducing the gene encoding engineered PHA synthase with acquired LA-polymerizing activity [20, 21]. They thus achieved the one-step biosynthesis of a copolymer with 6 mol% of lactate and 94 mol% of 3-hydroxybutyrate units. This extremely important result represents a milestone towards the biological synthesis of PLA and confirms that the work is moving in the right direction, as mentioned in the section of 2.3. At present, the LA fraction in the copolyesters has been enriched up to 96 mol% [22], so the synthesis of homopolymers of LA represents a major goal. To that end, the current microbial cell factory ought to be improved with further evolved LA-polymerizing enzymes (LPE) and metabolic engineering-based optimization [23, 24]. Matsumura et al. likewise reported the lipase PC-catalyzed polymerization of cyclic diester-D,L-lactide at a temperature of 80-130 °C to yield PLA with molecular masses of up to 12,600 [25].

1.3. Toward an enhanced sustainable production

Three main issues have hindered widespread use of PHAs: (1) the high production cost compared to petroleum-based polymers with similar properties; (2) the inability to produce high-performance PHAs in substantial amounts; and (3) the difficulty in controlling the life cycle of PHAs, i.e., the control of their biodegradability and their effective chemical recycling.

To solve the former two issues, we have focused on the genetic engineering of PHAs metabolism, which will lead to the cost-effective biological production of PHAs and the improvement of their properties, such as molecular mass and monomer composition. In particular, protein engineering of PHA synthase can improve both PHA production efficiency and the properties of the generated polymer because PHA synthase plays a central role in PHA biosynthesis [26]. Here we would like to highlight the current special topic on the biosynthesis of new PHA polymers incorporating unusual monomer units such as LA by PHA synthase engineering. Further, gene cloning and expression in plants has created new possibilities of using photosynthesis to convert atmospheric CO_2 directly into PHA, in hopes of reducing production cost in the future.

In addition, to solve the latter issue, we have also focused on the engineering of PHB depolymerases. PHB is the most common form of PHAs. In natural environment, the microbial and enzymatic degradation of PHB is an important first step in the PHB recycling process. However, PHB degradation depends on the surrounding conditions and proceeds on the order of a few months in anaerobic sewage or a few years in seawater [13]. Such PHB degradation process is undesirable from the standpoint of the efficient use of biomass resources. To overcome this issue, chemical recycling using spent PHB materials as recyclable monomer-concentrated resources is rapidly gaining importance due to its high degradation rate [27]. In addition, as chemical recycling is cost-efficient and has low CO_2 emissions, it has great potential as a low-cost and environmentally compatible process. PHB monomerization, the first step in chemical recycling, is currently carried out via a thermal decomposition process. However, this chemical recycling method presents some drawbacks, such as racemization of the decomposed products, high reaction temperature, and contamination with residual metal catalysts [28-31]. As one of the solutions, the development of alternative PHB monomerization methods that use such enzymes as PHB depolymerases is highly awaited because those methods do not produce undesirable byproducts, have high enantio- and regioselectivities, and can be performed at moderate temperatures [32, 33]. Moreover, as the efficient use of biocatalysts requires suitable enzymes with high activity and stability under process conditions, the desired substrate selectivity, and high enantioselectivity, the improvement of PHB depolymerases is expected to result in the construction of an effective PHB chemical recycling system. In this chapter we will also provide some case studies on protein engineering of PHB depolymerase based on domain structure-based and random mutagenesis approaches.

2. Protein engineering of PHA synthases

2.1. Biochemical properties and engineering concepts of PHA synthases

PHA synthases catalyze the polymerization reaction of hydroxyalkanoate (HA) to polymer PHA. The monomer substrates of PHA synthase are mainly 3HA-CoAs with various side-chain lengths, and only R-enantiomer HA-CoAs are accepted for polymerization by synthase [34]. Over 60 different PHA synthases have been classified into four types based on their substrate specificities and subunit compositions of enzymes (Table 1) [35].

Type	Subunit(s)	Representative species	Substrate specificity
I	PhaC	*Ralstonia eutropha*	C3 - C5
		Aeromonas caviae	C3 – C7
II	PhaC	*Pseudomonas aeruginosa*	C6 – C14
		Pseudomonas sp 61-3	C3 – C12
III	PhaC - PhaE	*Allochromatium vinosum*	C3 – C5
IV	PhaC - PhaR	*Bacillus megaterium*	

T. Hiraishi and S. Taguchi

Table 1. The four classes of PHA synthases

Type I and type II PHA synthases consist of single subunits (PhaC). Type I PHA synthases, represented by *Ralstonia eutropha* enzyme, mainly polymerize SCL-monomers (C3–C5), whereas type II PHA synthases, represented by *Pseudomonas oleovorans* enzyme, polymerize MCL-monomers (C6–C20). Type III PHA synthases, represented by *Allochromatium vinosum* enzyme, consist of two hetero-subunits (PhaC and PhaE). PhaC subunits of type III synthase are smaller than those of type I and II synthases, but possess catalytic residues. Like the type I synthases, these PHA synthases prefer to polymerize SCL-monomers (C3–C5). Type IV PHA synthases, represented by *Bacillus megaterium*, are similar to the type III PHA synthases with respect to possessing two subunits. However, unlike the PhaE of type III PHA synthases, a smaller protein designated as PhaR is required for full activity expression of type IV PhaC.

The lack of a suitable structural model for any PHA synthase has limited attempts to improve the activity and to alter the substrate specificity of these enzymes in "irrational" manners, such as random mutagenesis and gene shuffling [36, 37]. Generally, natural diversity provides us with attractive starting materials for artificial evolution as it represents functionalized sequence spaces to some extent. A tremendous population (over 60 species) of randomly screened PHA producing bacteria suggests that attractive prototype enzymes for molecular breeding would exist. Among them, enzyme evolution approach has been applied to the following type I and type II PHA synthases derived from some bacteria.

2.2. Activity improvement and substrate specificity alteration of PHA synthases

2.2.1. Application to type I PHA synthases

A pioneering study that established methods for protein engineering PHA synthase initiated in 2001 using the best-studied enzyme, the *R. eutropha* PHA synthase [38]. *In vitro* evolutionary program was firstly constructed by coupling an error-prone PCR-mediated point mutagenesis with the plate-based high-throughput screening method to generate mutants with acquired beneficial functions [38]. A mutant library of the *R. eutropha* PHA synthase gene was prepared by colony formation of transformant cells of *Escherichia coli*. It should be noted to meet a good correlation between the change in PHB accumulation resulting from the introduction of mutations into the *R. eutropha* PHA synthase gene and the change in the enzymatic activity of the mutants. To gain the mutants with increased activity, multi-step mutations, including an activity loss and an intragenic suppression-type activity reversion were attempted [39]. The mutant enzymes were once identified by primary mutation analysis, a secondary round of mutation was used to evolve these enzymes to proteins with better characteristics than the wild-type enzyme. As a result, through this intragenic suppression-type mutagenesis, an increased specific activity towards 3HB-CoA by 2.4-fold compared to the wild-type enzyme was acquired by a mutation of Phe420Ser (F420S) in a type I PHA synthase [39].

As a next case, screened beneficial mutation, Gly4Asp (G4D), exhibited higher levels of protein accumulation and PHB production compared to the recombinant *E. coli* strain harboring the wild-type PHA synthase [40]. As for intragenic suppression-type mutagenesis, second-site reversion is dependent or independent of primary mutation in the activity. Secondary mutations of F420S and G4D are the latter cases, being independent of primary mutation. Subsequently, site-specific saturation mutagenesis was also performed on the codon encoding the G4 residue of the *R. eutropha* PHA synthase and many substitutions resulted in much higher PHB content as well as higher molecular masses of the polymers [41].

Aeromonas caviae (*punctata*) PHA synthase is unique among type I PHA synthases since it can synthesize not only PHB homopolymer but also random copolyesters of 3HB and 3-hydroxyhexanoate (3HHx). Kichise et al. performed the first successful *in vitro* molecular evolution experiments on PHA synthase from *A. caviae* by targeting to the limited region of the enzyme [42]. Two evolvants exhibited increased activity towards 3HB-CoA of 56% and 21%, respectively, compared to the wild-type enzyme by *in vitro* assays. These mutations led to enhanced accumulation (up to 6.5-fold higher than the wild-type enzyme) of P(3HB-*co*-3HHx) and increases in the 3HHx molar fraction (16-18 mol% compared to 10 mol% of the wild-type PHA synthase) in recombinant *E. coli* strains grown on dodecanoate. As an extended study, a combination of these two beneficial mutations (N149S/D171G) synergistically altered enzymatic properties, leading to synthesis of PHA copolymers with enhanced 3HA fraction and increased molecular mass from in the recombinant *R. eutropha* [43]. In a separate study, *A. caviae* PHA synthase was engineered *in vivo* using the mutator strain *E. coli* which has a 5,000-fold higher mutation rate than wild-type *E. coli*, and mutants were again screened for enhanced PHB accumulation in recombinant *E. coli* [44]. Also, mutants synthesized PHAs with increased molecular mass, but in contrast to the previous study, the 3HHx fraction was only slightly different from wild-type composition.

Junction site for interconnection of heterogeneous enzymes based on the predicted secondary structures allowed chimeragenesis of the PHA synthase from *R. eutropha* with the partner PHA synthase from *A. caviae*. Successfully obtained chimera-mutant exhibited improved activity increase and expanded substrate specificity compared to the original enzymes [45]. As for PHA synthases, directed evolution studies have thus progressed through advancements from random approach to much more systematic approaches such as chimera-genesis, recombination and shuffling.

2.2.2. Application to type II PHA synthases

Contrasted with the type I PHA synthases, type II PHA synthases typically have substrate specificity towards MCL-3HA-CoA substrates but relatively poor substrate specificity towards SCL-3HA-CoA substrates like 3HB-CoA. An exception to this is the type II PHA synthase of *Pseudomonas* sp. 61-3 with significant substrate specificity towards the 3HB-CoA (Table 1). In the landmark study by Takase *et al*, the *in vitro* evolutionary technique was applied to the PhaC1 PHA synthase from *Pseudomonas* sp. 61-3 to increase the activity towards 3HB-CoA monomers [46]. Substitutions at two amino acid residues, Ser325 and Gln481 were found to dramatically effect the production of PHB homopolymer in recombinant *E. coli* with glucose as the carbon source. The codons for these amino acids were subjected to site-specific saturation mutagenesis and several individual substitutions were found that could dramatically increase the level of PHB production. These mutations were combined as double mutants to further increase the level of PHB production (340 - 400-fold higher than the wild-type enzyme) [46]. The changes in the *in vivo* produced P(3HB-*co*-3HA) copolymer molar compositions correlated well with the *in vitro* biochemical data of the substrate specificity and activity of the enzymes and represents one of the most well-rounded studies to date [47].

The findings obtained in these studies for the type II PHA synthase would be very useful for evaluating a similar evolution strategy to the other types of PHA synthases based on the amino acid sequence alignment of the PHA synthases. For example, position 481 in PhaC1 PHA synthase from *Pseudomonas* sp. 61-3 was found to be one of the residues determining substrate specificity of the enzyme, as described above. Interestingly, the amino acid residues corresponding to the position of this enzyme are conserved within each type of PHA synthases; Ala for type I, Gln for type II, Gly for type III and Ser for type IV enzymes. Thus, the effects of mutating the highly conserved alanine (Ala510) of the *R. eutropha* PHA synthase (corresponding to the position 481 in *Pseudomonas* sp. 61-3 PhaC1) were analyzed via site-specific saturation mutagenesis. Mutations at Ala510 were found to affect the substrate specificity of the *R. eutropha* PHA synthase, allowing slightly higher 3HA incorporation compared to the wild-type PHA synthase in *R. eutropha* PHB⁻4 (PHA negative mutant) [48].

The other two beneficial positions, Glu130 [49] and Ser477 [50], were also identified through the *in vitro* evolution screening. As illustrated in Figure 2 (A) and (B), "mutation scrambling" among four beneficial positions (130, 325, 477, 481) for activity increase, change in substrate specificity, and regulation of polymer molecular mass would further create new super-enzymes. Most recently, a possible mechanistic model for PHA polymerization has been proposed on the basis of the accumulated evolutionary studies [51]. Furthermore, the useful evolvants obtained through the systematic enzyme evolution have been supplied to other organisms including plants [52, 53]. The impacts of these reports prompted the other research groups to apply directed evolution to the individual PHA synthases of interests [54-56].

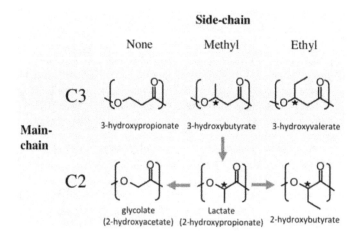

T. Hiraishi and S. Taguchi

Figure 2. Structural diversities in main-chain and side-chain of PHA back bone which can be recognized by natural and artificially evolved PHA synthases. Asterisks indicate the chiral center in monomer units of PHA.

2.3. Engineering of lactate-polymerizing enzyme (LPE) from PHA synthases

The pioneering work on the exploration of LA-polymerizing activity by PHA synthases was reported by Valentin et al. [57]. In that attempt, the PLA biosynthesis was carried out by monitoring the activity of PHA synthases towards synthetic LA-CoAs (*R* and *S* enantiomers). Several PHA synthases were evaluated for LA-polymerizing activities and a class III PHA synthase from *Allochromatium vinosum* exhibited a weak CoA releasing activity [57]. In a similar report, Yuan et al. reported in detail the activity of *A. vinosum* PHA synthase towards (*R*)-LA-CoA [58]. Unfortunately, in either case, polymerization was not observed/was negligible, suggesting that PHA synthase could hydrolyze CoA ester to release CoA but not progress from there with polymerization to form a polymer.

In this context, Taguchi et al. formally reported the first prototype LPE in the year 2008 as a PHA synthase with an acquired LA-polymerizing activity through *in vitro* and *in vivo* experiments [20]. The first clue of LA-polymerizing activity was obtained through a water-organic solvent two-phase *in vitro* system [20, 59]. The activity towards LA-CoA was tested in the absence or presence of 3HB-CoA using representative PHA synthases belonging to the four classes of natural PHA synthases together with three engineered (PhaC1$_{Ps}$ mutants) from *Pseudomonas* sp. 61-3. The engineered PHA synthases were two single mutants[Ser325Thr (ST) and Gln481Lys (QK)] and one double mutant carrying the two mutations (STQK). The two mutants were selected based on their improved activity and/or broader substrate specificity [36, 37]. The natural synthases and mutants did not form a clear-polymer with LA-CoA alone but did with 3HB-CoA. However, when LA-CoA was supplied together with 3HB-CoA, one mutant, PhaC1$_{Ps}$(STQK) clearly exhibited a polymer-like precipitation. Subsequently, the analysis of the precipitant revealed that the precipitant consisted of 36 mol% of the LA unit. Therefore, this was the first report ever of a PHA synthase with ability to incorporate LA unit to form P(LA-*co*-3HB).

The finding that PhaC1$_{Ps}$(STQK) could polymerize LA was a demonstration of evolutionary engineering as a powerful tool for the generation of biocatalysts with desired properties. By demonstrating the *in vitro* activity of PhaC1$_{Ps}$(STQK) towards LA-CoA, it was presumed that heterologous expression of this LPE could result into an *in vivo* synthesis of LA-based poly-esters thus creating microbial factories for LA-based polyesters synthesis.

In a subsequent study, based on the improved activity of a point mutation at position 420 (F420S) of a type I PHA synthase (PhaC$_{Re}$) from *R. eutropha* (Taguchi et al., 2002) [39], the same mutation was introduced into the ancestral LPE [PhaC1$_{Ps}$(STQK) from *Pseudomonas* sp. 61-3] to create a triple mutant of LPE with S325T and Q481K along with a new mutation, F392S which corresponds to F420S of PhaC$_{Re}$ [60]. When the new further engineered LPE [PhaC1$_{Ps}$ (STQKFS)] was expressed in *E. coli*, a copolymer with 45 mol% LA and polymer content of 62 wt% was synthesized in comparison with P(26 mol% LA-*co*-3HB) obtained with the prototype LPE, PhaC1$_{Ps}$(STQK) under aerobic culture conditions. Additionally, the cells harboring PhaC1$_{Ps}$ (STQKFS) synthesized P(LA-*co*-3HB) with 62 mol% LA with polymer content of 12 wt %. During the same study, saturation mutagenesis of LPE at the same site (position 392) yielded mutants that gave varying LA fractions in the copolymers however; F392S was superior to the other mutants in incorporating LA. This study demonstrated the effectiveness of enzyme engineering of the LPE towards two directions; there was improved LA incorporation and polymer yield improvement for both aerobic and anaerobic culture conditions [60]. Furthermore, it will be interesting to note that copolymers incorporating 2-hydroxy acids (2HAs) such as 2-hydroxybutyrate [61] and glycolate (Matsumoto et al., 2011) [62] may lead to copolymers with novel properties (Figure 3). This expansion of PHA synthase to 2HAs-polymerizing enzymes has extensively prompted us to create further new enzymes with acquired activities toward new unusual monomer substrates, consequently create new polymers.

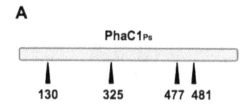

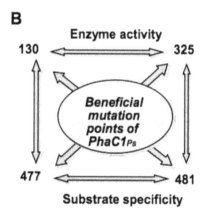

T. Hiraishi and S. Taguchi

Figure 3. Functional mapping of beneficial positions (A) and relationships among the residues related to enzymatic activity and substrate specificity of PHA synthase from *Pseudomonas* sp. 61-3 (B).

Regarding the reports on LPE, the following several studies have been published [63, 64]. Currently, the best-studied PHA synthase from *R. eutropha* has been successfully engineered to LPE by only single mutations at beneficial position corresponding to the position 481 of *Pseudomonas* sp. 61-3 PHA synthase [65]. This implies the functional compatibility between PHA members also for acquiring LPE activity. In the prospect, advanced types of LPE will be supplied by artificial evolution of the prototype LPE as well as exploration of natural PHA synthases with potentially possessing new substrate specificities such as LA-polymerizing activity.

3. Protein engineering of PHB depolymerase

3.1. Biochemical and genetic properties of PHB depolymerases

A number of PHA depolymerases have been purified from diverse PHA-degrading microorganisms and characterized [9, 10, 12]. As described earlier, depending on the substrates and localization of PHA depolymerases, PHA depolymerases are grouped generally into four families: PHA depolymerases degrading the native intracellular granules (i-PHA$_{MCL}$ depolymerases and i-PHA$_{SCL}$ (i-PHB) depolymerases) and PHA depolymerases degrading the denatured extracellular PHA granules (e-PHA$_{MCL}$ depolymerases and e-PHA$_{SCL}$ (e-PHB) depolymerases). To date, the genes of about 30 PHA depolymerases with experimentally verified PHA depolymerase activity have been identified. On the basis of their sequence similarity, the PHA Depolymerase Engineering Database has been established as a tool for systematic analysis of PHA depolymerase family [66].

Among the PHA depolymerases, multi-domain e-PHB depolymerases have been extensively examined [9]. The multi-domain e-PHB depolymerases generally have a domain structure consisting of a catalytic domain (CD) at N-terminus, a substrate-binding domain (SBD) at C-terminus, and a linker region connecting the two domains, while e-PHB depolymerases from *Penicillium funiculosum* (PhaZ$_{Pfu}$) and PhaZ7 from *Paucimonas lemoignei* (PhaZ7$_{Ple}$) have emerged as two exceptions (single-domain e-PHB depolymerases) [9, 67-69]. Genetic analysis also shows that e-PHB depolymerases contain a lipase box pentapeptide [Gly-X$_1$-Ser-X$_2$-Gly] as an active residue, indicating that these enzymes are one of the serine hydrolases. As an example, the domain structure of e-PHB depolymerase from *Ralstonia pickettii* T1 (PhaZ$_{RpiT1}$) is illustrated in Figure 4(A). Such domain structure has been found in many biopolymer-degrading enzymes, such as cellulase, xylanase, and chitinase, which are capable of hydrolyzing water-insoluble polysaccharides [70-73]. The enzymatic degradation of PHB by the multi-domain e-PHB depolymerases is considered to proceed via a two-step reaction at the solid-liquid interface, as shown in Figure 4(B). The e-PHB depolymerase approaches and adheres to the PHB surface via SBD, followed by hydrolysis of the polymer chain by CD. Accordingly, it is considered that elucidation of the mechanisms of enzyme adsorption and enzymatic hydrolysis will contribute to the development of new PHB polymer materials with the desired environmental stability and biodegradability as well as the development of improved e-PHB depolymerases that can be used to effectively recycle PHB materials.

From a biological viewpoint, the structure-function relationship of maluti-domain e-PHB depolymerases has been studied extensively, and several mutants were designed to analyze the function of each domain, in particular, SBD. Using a truncated multi-domain e-PHB depolymerases, Behrends et al., Nojiri and Saito, and our group revealed that the C-terminal domain is essential for PHB-specific binding [74-76]. Further, Nojiri and Saito genetically prepared many mutants of PhaZ$_{RpiT1}$ in various forms such as inversions, chimeras, and fusion

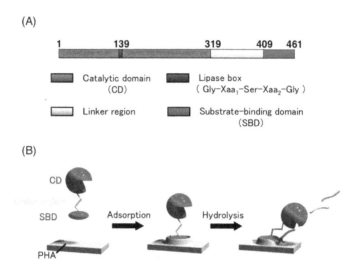

(A)

1   139   319   409   461

▨ Catalytic domain   ▨ Lipase box
   (CD)   (Gly–Xaa$_1$–Ser–Xaa$_2$–Gly)

▨ Linker region   ▨ Substrate–binding domain
                               (SBD)

(B)

CD

SBD   Adsorption   Hydrolysis

PHA

T. Hiraishi and S. Taguchi

Figure 4. A) Domain structure of e-PHB depolymerase from *Ralstonia pickettii* T1 (PhaZ*RpiT1*). (B) Schematic illustration of the enzymatic degradation of PHA by e-PHB depolymerase.

to extra linker domains, and demonstrated that its SBD organization also influences the PHB degradation but not water-soluble substrates. Doi and co-workers prepared fusion proteins of SBDs of several PHB depolymerases with glutathione-S-transferase [77-81], and demonstrated specific interactions based on molecular recognition between SBD and polyester surface.

3.2. Effects of chemical and solid-state structures and surface properties of PHAs on enzymatic degradation

Chemical structures of PHAs have influence on their enzymatic hydrolysis by multi-domain e-PHB depolymerases. Various types of PHAs including racemic PHA [82-89] and 3HA oligomers [90, 91], PHAs with different main- and side-chain lengths (Kasuya et al., 1997) [77], and random copolymers of (R)-3HB with various hydroxyalkanoate units [92-95] have been synthesized to examine their enzymatic degradation by a variety of e-PHB depolymerases. For instance, Abe et al. proposed a schematic model of the enzymatic cleavage of the PHA chain by PhaZ$_{RpiT1}$ (Figure 5), in which its active site can recognize at least three neighboring monomer units with a certain degree of difference in main-chain length [93]. Besides the chemical structure, the solid-state structure and surface properties of PHAs also influence the enzymatic hydrolysis. For example, the amorphous regions in PHA materials are preferentially hydrolyzed, followed by the hydrolysis of crystalline regions as a rate-limiting step in the enzymatic degradation process [96, 97]. Further, the enzymatic degradation rate of PHA materials decreases with increasing crystallinity, crystal size, and regularity of the chain packing state. In addition, Abe and co-workers demonstrated using proteinase K that the change in the surface properties of PLA film induced by end-capping with alkyl ester groups (carbon numbers 12 to 14) leads to a decrease in their enzymatic degradation rates [98, 99].

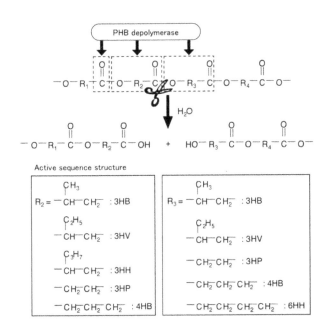

Figure 5. Schematic model of enzymatic cleavage of an ester bond in various sequences by PHB depolymerase.

To investigate the influence of the chemical structure or surface properties of polymer on enzymatic adsorption at nano-level sensitivity, several studies using quartz crystal microbalance (QCM) and atomic force microscopy (AFM) have been performed. Yamashita et al. investigated the $PhaZ_{RpiT1}$ adsorption to the film surface of several polymers including polyethylene, polystyrene and PHA using the QCM technique, and found that the enzyme showed adsorption specificity for PHA [100-102]. In addition, AFM analysis of $PhaZ_{RpiT1}$ on polyester surface has revealed that small ridges are formed around the enzyme molecule due to movement of some polyester chains at the adsorption area, suggesting that a strong chemical interaction exists between the enzyme and the polyester chains [102, 103]. Furthermore, AFM analysis of interaction between PHB single crystal and a hydrolytic-activity-disrupted $PhaZ_{RpiT1}$ mutant has demonstrated that its SBD disturbs the molecular packing of PHB polymer chains, resulting in fragmentation of the PHB single crystal [104]. Taking these findings into consideration, the specific adsorption of PHB depolymerase to the PHB surface probably involves both the adsorption of the enzyme to the surface and the non-hydrolytic disruption of the substrate to promote PHB degradation. Recently, we have developed the AFM technique by using an AFM tip modified with SBD protein to evaluate the interaction between the SBD molecule and the PHB surface at the molecular level. Through this, it has been shown that the adsorption force of one SBD molecule to the PHB surface is approximately 100 pN [105, 106].

3.3. Analysis of polymer binding ability of e-PHB depolymerase using directed evolution technique

The structural aspects of an enzyme generally provide crucial information about the interaction between the enzyme and its ligand. Some researchers have reported the tertiary structures of polymer-degrading enzymes, such as glycoside hydrolases and single-domain e-PHB depolymerases, and proposed an interaction model between the enzymes and the polymer surfaces [68, 107-109]. However, because of the paucity of information about the 3D structures of multi-domain e-PHB depolymerases, there are few insights into which and how amino acid residues in their SBD are involved in the enzyme adsorption to PHB surface.

Directed evolution is a useful and powerful tool to explore, manipulate, and optimize the properties of an enzyme as no information on the tertiary structure of the enzyme is required and new and unexpected beneficial mutations can be discovered [110-112]. Random mutagenesis via error-prone PCR (epPCR) and DNA recombination are widely used approaches to generate a large mutant pool and screen for the desired characteristics [113, 114]. Using those approaches, many enzymes with improved substrate specificity, catalytic activity, thermostability, or solubility were obtained [115]. Further, analysis of the effects of mutations could also provide useful information for the improvement of enzyme function.

To improve e-PHB depolymerases, it is important to understand the mechanisms underlying its adsorption and hydrolysis, such as which and how amino acid residues participate in the catalytic process. To clarify this issue, we have investigated the interaction between PhaZ$_{\text{RpiT1}}$ and PHB surface by a combination of PCR random mutagenesis targeted to only SBD and an *in vivo* screening system as shown in Figure 6(A) [116]. In the analysis of recombinants showing low PHB-degrading activity, Ser410, Tyr412, Val415, Tyr428, Ser432, Leu441, Tyr443, Ser445, Ala448, Tyr455, and Val457 were replaced with other residues having hydropathy indices opposite to theirs at high frequency (Figure 6(B)). The results suggested that PhaZ$_{\text{RpiT1}}$ adsorbs to the PHB surface not only via the formation of hydrogen bonds between hydroxyl groups of Ser at these positions of the enzyme and carbonyl groups in the PHB polymer, but also via the hydrophobic interaction between hydrophobic residues at abovementioned positions and methyl groups in the PHB polymer.

Nevertheless, because only little knowledge was obtained on the biochemistry and kinetics of the purified mutant enzymes, the roles of these amino acids (Ser410, Tyr412, Val415, Tyr428, Ser432, Leu441, Tyr443, Ser445, Ala448, Tyr455, and Val457) and their contributions to the enzymatic activity remain poorly understood, resulting in little information to develop e-PHB depolymerases. Among these positions, Leu441, Tyr443, and Ser445 were predicted to form a β-sheet structure to orient in the same direction as shown in Figure 6(B). As polymer-degrading enzymes generally align their amino acid residues in a plane to interact with polymer surfaces,

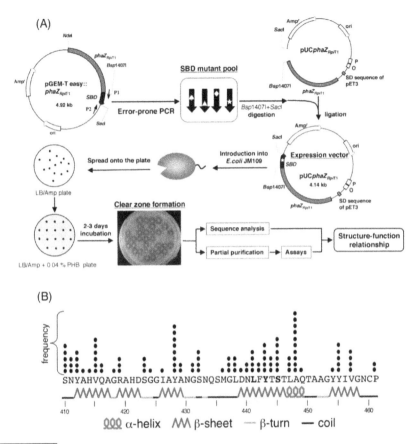

T. Hiraishi and S. Taguchi

Figure 6. A) *In vivo* assay system for assessment of mutational effects of the substrate-binding domain of PhaZ$_{RpiT1}$ on PHB degradation. Schematic flow diagram of the system is illustrated. This system is composed of PCR-mediated random mutagenesis in the substrate-binding domain region of PhaZ$_{RpiT1}$ gene, preparation of mutant library, primary plate assay of PHB degradation (clear-zone formation), nucleotide sequencing and PHB degrading and adsorbing assays of partially purified mutant enzymes. (B) Positions and frequencies of PCR-mediated single mutations in the region coding for SBD of PhaZ$_{RpiT1}$, together with its predicted secondary structure.

these three residues in PhaZ$_{RpiT1}$ may interact directly with the PHB surface. Since the hydropathy indices of such mutations as L441H (replacement of Leu441 with His), Y443H (replacement of Tyr443 with His), and S445C (replacement of Ser445 with Cys) dramatically changed among the mutations at these positions, their PHB-binding and -degrading properties were examined in detail [117]. Functional analyses of the purified L441H, Y443H, and S445C enzymes indicated that these mutations had no influence on their structures and their ability to cleave the ester bond, while their PHB-degrading activity differed from that of the wild type. Kinetic analysis of PHB degradation by the mutants suggested that the hydrophobic residues at these positions are important for the enzyme adsorption to the PHB surface, and may more effectively disrupt the PHB surface to enhance the hydrolysis of PHB polymer chains than the wild-type enzyme. Further, surface plasmon resonance (SPR) analysis revealed that these substitutions mentioned above altered the association phase rather than the dissociation phase in the enzyme adsorption to the polymer surface.

Recently, Hisano et al. determined the crystal structure of PhaZ$_{Pfu}$ and proposed that hydrophobic residues, including Tyr, Leu, Ile, and Val, contribute to adsorption to the PHB surface, and that hydrophilic residues (Ser and Asn) located around the mouth of the enzyme crevice may also contribute to the affinity of the enzyme for PHB [68]. Jendrossek group determined PhaZ7$_{Ple}$ crystal structure and demonstrated that the enzyme was enriched in hydrophobic amino acids including eight tyrosine residues [108]. All tyrosine residues (Tyr103, Tyr105, Tyr172, Tyr173, Tyr189, Tyr190, Tyr203, and Tyr204), which are located at the surface of PhaZ7$_{Ple}$ but are far from the active site (Ser136), were changed to alanine or serine and the substitution effects were examined [118]. It turned out that mutation of Tyr105, Tyr189 or Tyr190 resulted in reduced PHB-degrading activity and in occurrence of a lag phase of the depolymerase reaction, indicating that these residues are possibly involved in the enzyme adsorption. Similar results have been obtained for the e-PHA$_{MCL}$ depolymerase of *Pseudomonas fluorescens* GK13 by Jendrossek et al. [119]. They reported that several hydrophobic amino acids (Leu15, Val20, Ile26, Phe50, Phe63, Tyr143 and Val198) were identified to be involved in interaction between the enzyme and poly(3-hydroxyoctanoate) substrate surface. This finding was supported with the recent study by Ihssen et al. (2009) [120].

3.4. Improvement in SBD function of PhaZ$_{RpiT1}$

The above-mentioned findings imply that PHB binding ability of PhaZ$_{RpiT1}$ can be improved by substituting a hydrophilic residue with a hydrophobic one at the positions of 441, 443 and 445. Tyr at position 443 was targeted for substitution with a more highly hydrophobic amino acid residue because its hydrophobicity shows medium to high degree compared to those of general naturally occurring amino acid residues [121].

Table 2 shows the hydrophobicity, the potential for β-sheet formation, and the volume of 20 common amino acid residues [122-124]. In this table, the properties of the original amino acid residue are colored blue and the desirable characteristics of the amino acid residues are colored orange, respectively. In the design of a mutant enzyme with an amino acid substitution at this position, the following factors were taken into consideration: (1) to achieve higher hydrophobicity than the original residue, (2) to retain the β-sheet structure, and (3) to change as little as possible the volume of the amino acid residue after the substitution. As a result, the substitution of Tyr443 with Phe (Y443F) was considered to be appropriate. Analysis of the purified Y443F enzyme indicated that the mutation had no influence on the structure and the ester bond cleavage activity, while this mutant had higher PHB degradation activity than the wild type. Thus, this finding supports our previous assumption and indicates the importance of highly hydrophobic residues at these positions for PHB degradation.

amino acid residue	hydrophobicity[a]	Pb[b]	volume[c]
Ile	4.5	1.60h	100.1
Val	4.2	1.70h	83.9
Leu	3.8	1.30h	100.1
Phe	2.8	1.38h	113.9
Cys	2.5	1.19h	65.1
Met	1.9	1.05h	97.7
Ala	1.8	0.83i	53.2
Gly	-0.4	0.75b	36.1
Thr	-0.7	1.19h	69.7
Ser	-0.8	0.75b	53.4
Trp	-0.9	1.37h	136.7
Tyr(wild type)	-1.3	1.47h	116.2
Pro	-1.6	0.55b	73.6
His	-3.2	0.87i	91.9
Asn	-3.5	0.89i	70.6
Gln	-3.5	1.10h	86.3
Asp	-3.5	0.54b	66.7
Glu	-3.5	0.37b	83.0
Lys	-3.9	0.74b	101.1
Arg	-4.5	0.93i	104.1

T. Hiraishi and S. Taguchi

[a] J. Kyte and R. F. Doolittle, 1982 [122].

[b] P. Y. Chou and G. D. Fasman, 1978 [123]. Pb: potential for β-sheet formation ; h: former ; i: indifferent ; b: breaker.

[c] A. A. Zamyatnin, 1972 [124].

Table 2. Hydrophobicity, potential for β-sheet formation, and volume of amino acid residues

3.5. Cell surface display system for protein engineering of PhaZ$_{RpiT1}$

Cell surface display is a valuable technique for the expression of peptides or proteins on the surface of bacteria and yeasts by fusion with the appropriate anchoring motifs [125]. Therefore, the cell surface display of functional and useful peptides and proteins, such as enzymes, receptors, and antigens, has become an increasingly used strategy in various applications, including whole-cell biocatalysts and bioabsorbents, live vaccine development, antibody production, and peptide library screening. In addition, this method is very useful for enzyme library screening because the displayed protein is accessible to the external environment and thus, is able to interact with substrates easily, allowing the screening of large libraries [126].

A variety of surface anchoring motifs, including outer membrane proteins, lipoproteins, autotransporters, subunits of surface appendages, and S-layer proteins, have been employed to achieve the display systems [125, 127, 128]. We used the OprI anchoring motif for the functional display of PhaZ$_{RpiT1}$ on *Escherichia coli* cell surface [129]. The displayed enzyme retained its intrinsic characteristics, that is, hydrolytic activity for *p*-nitrophenyl butyrate (pNPC4) and the ability to adsorb to and degrade PHB, indicating that the engineered *E. coli* can be used in the form of a whole-cell biocatalyst by overcoming the uptake limitation of such substrates as insoluble PHB. These findings also indicate that the whole-cell catalyst is a promising and suitable tool to screen for mutant PhaZ$_{RpiT1}$ with enhanced catalytic activity.

3.6. Protein engineering of CD region of PhaZ$_{RpiT1}$ using cell surface display system

In contrast to SBD, there is little knowledge on the CD of PhaZ$_{RpiT1}$, such that which and how amino acid residues in the CD contribute to the enzymatic activity remain poorly understood, and this has resulted in the lack of information for the improvement of the CD function of PhaZ$_{RpiT1}$. The CD of PhaZ$_{RpiT1}$ was targeted for the directed evolution, employing random mutagenesis and DNA recombination to enhance its ester bond cleavage ability (Figure 7) [130]. The mutant genes generated from these reactions were expressed as surface-displayed enzymes, and the mutant enzymes were screened through a high-throughput system using pNPC4, a water-soluble substrate. As a result, clones displaying mutant enzymes with a 4- to 8-fold increase in pNPC4 hydrolysis activity were obtained in comparison with those display-ing the wild type. This result was roughly consistent with the results of pNPC4 hydrolysis using purified enzymes with the unfused and undisplayed forms, concluding that the current screening system is feasible and effective for the search of improved enzymes.

As the aliphatic part in pNPC4 is similar to the monomer unit in PHB polymer chain and pNPC4 is generally used as a model substrate, changes in pNPCn hydrolysis rates by the purified mutant enzymes as a function of the chain length of the aliphatic part in *p*-nitrophenyl esters (pNPCn, n=2 to 6) can provide the information regarding the substrate recognition of the enzyme. The results of pNPCn hydrolysis by the mutants demonstrated that the elevation on their pNPCn hydrolysis activity for each pNPCn substrate occurred. DNA sequencing showed that eight improved mutant enzymes contained N285D or N285Y mutations. As beneficial mutations are accumulated and deleterious mutations are simultaneously removed from the improved mutants through DNA recombination procedures [131], the N285D and N285Y mutations found here are probably beneficial for pNPCn hydrolysis. Kinetic studies revealed that the increase in catalytic efficiency for pNPCn hydrolysis by the mutant enzymes is attributed to the high V_{max} values.

As opposed to pNPCn hydrolysis by the N285D and N285Y mutant enzymes, their PHB degradation rates were slower than that of the wild-type enzyme, indicating that these mutations are unfavorable for PHB degradation. The kinetics of PHB degradation demon-strated that the N285D and N285Y mutations lowered the hydrolysis activity for the PHB

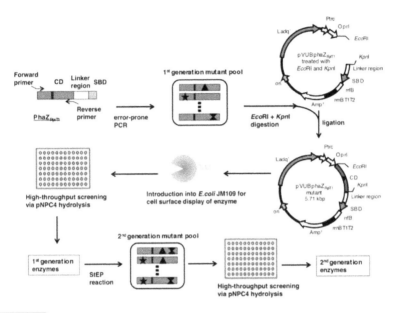

T. Hiraishi and S. Taguchi

Figure 7. Directed evolution targeted at the catalytic domain (CD) of PhaZ$_{RpiT1}$ using the *in vivo* screening system in the cell surface display system. A schematic diagram of the mutational effects analysis is illustrated. This system consists of random mutagenesis by error-prone PCR in the CD of PhaZR$_{piT1}$, cell surface display of enzyme, high-throughput micro-plate screening via *p*-nitrophenyl butyrate (pNPC4) hydrolysis, staggered extension process (StEP), and nucleotide sequencing.

polymer chain compared to the wild-type enzyme despite retention of the binding activity for the PHB polymer surface.

3.7. Proposed models of the active site in e-PHB depolymerases

The correct orientation of a PHB polymer chain to the active site is necessary to realize effective PHB degradation by e-PHB depolymerase. Hisano et al. have determined the crystal structures of PhaZ$_{Pfu}$-3HB trimer complex as well as PhaZ$_{Pfu}$ enzyme alone [68]. In the PhaZ$_{Pfu}$-3HB trimer complex, 3HB trimer binds to the crevice with its carbonyl terminus towards the catalytic residues (Figure 8(A)). From the structural insight gained from PhaZ$_{Pfu}$, they proposed the mechanism of action of PhaZ$_{Pfu}$. Figure 8(B) shows the location of the catalytic residues and the interaction between PHB polymer chain and the residues in the subsite of the active site of PhaZ$_{Pfu}$. In their model, Ser39 participates in the nucleophilic attack of the carbonyl carbon of a PHB chain, resulting in the formation of a covalent acyl-enzyme intermediate followed by the hydrolysis by an activated water molecule. The nucleophilicity of the hydroxyl group of Ser39 is enhanced by the His155-Asp121 hydrogen bonding system.

(A)

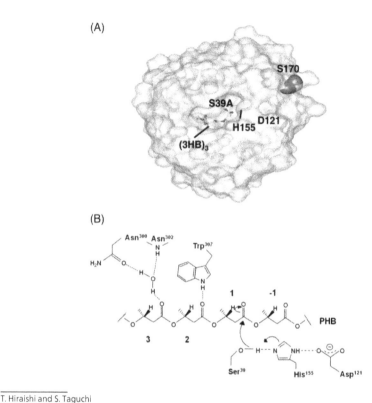

(B)

Figure 8. A) Molecular surface representation of PhaZ$_{Pfu}$. 3HB trimer in the crevice is shown as a ball and stick model. The positions of catalytic triad residues (S39A, D121, and H155) (cyan), as well as residue S170 (color-coded according to molecular species) are indicated. (B) Proposed model of the active site in PhaZ$_{Pfu}$ by Hisano et al. (Hisano et al. 2006).

For PhaZ$_{RpiT1}$, Bachmann and Seebach proposed that this enzyme has four subsites (2, 1, -1, and -2) in its active site, in which three of the subsites must be occupied by (*R*)-3-hydroxybutyrate (3HB) units for cleavage to occur at the center of the active site [90]. Homology modeling of PhaZ$_{RpiT1}$ using the SWISS-MODEL program based on the crystal structure of PhaZ$_{Pfu}$ (PDB accession no. 2d81A) was performed to speculate the possible localization of Asn285 in the active site. Figure 9(A) shows the homology modeling structure of PhaZ$_{RpiT1}$, in which the modeled residue range was positioned from 124 to 294. The residue Asn285 (color-coded according to molecular species) of PhaZ$_{RpiT1}$ is located at the mouth of the crevice and also located immediately above His273, which corresponds to His155 in subsite -1 of PhaZ$_{Pfu}$. However, Asn285 was positioned as if to cover the subsite -1 and to inhibit the substrate access. Taking the homology modeling results and the aforementioned information on the cleavage mechanism into consideration, we propose a simple schematic model for PhaZ$_{RpiT1}$, as shown in Figure 9(B). In this model, Ser139 participates in the nucleophilic attack of the carbonyl carbon of a PHB chain, and its nucleophilicity is enhanced by the His273-Asp121 hydrogen bonding system. The residue Asn285 is positioned relatively close to His273 located in subsite -1 as if to cover the subsite. The location of Asn285 in the subsite probably leads to the regulation of the recognition of substrate molecules, such as pNPCn and PHB polymer chain, possibly indirectly via conformational change. A similar situation has been described in lipases and PhaZ7$_{Ple}$ where activation via conformational change is required to uncover the active site [108, 109].

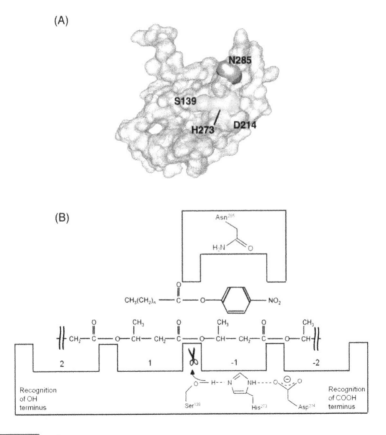

T. Hiraishi and S. Taguchi

Figure 9. A) Molecular surface representation of the homology model of PhaZ$_{RpiT1}$. The positions of catalytic triad residues (S139, D214, and H273) (cyan), as well as residue N285 (color-coded according to molecular species) are indicated. (B) Newly proposed schematic model of the active site in the CD of PhaZ$_{RpiT1}$.

4. Conclusion

This review describes the development of PHA synthases to synthesize the wide variety of custom-made bioplastics as well as PHB depolymerase with higher activity for PHB adsorption or pNPCn hydrolysis.

Bioplastics present a multitude of benefits as substitutes for conventional petroleum-based plastics. Among them, PHAs are one of the desirable alternatives to petrochemical-derived polymers because PHAs are produced directly from renewable resources completely by biological process and can be renewed over a relatively-short time. However, three main issues have hindered widespread use: the high production cost compared to petroleum-based polymers with similar properties; the inability to produce high-performance PHA polymers in substantial amounts; and the difficulty in controlling the life cycle of PHA polymers, i.e., the control of their biodegradability and their efficient recycling. Thus, with the development in recombinant DNA technology and high-throughput screening techniques, protein engineering methods and applications on the improvement of processes of bioplastic production as well as bioplastic degradation are becoming increasingly important and widespread.

The enzyme modification by protein engineering is an increasingly important scientific field. The well-known methods of rational design and directed evolution as well as new techniques including computational design, catalytic antibodies and mRNA display will be crucial for de novo design of enzymes. With recent advances in recombinant DNA technology tools including omics technologies and high-throughput screening facilities, improved methods for protein engineering will be available for easy modification or improvement of more enzymes for further specific applications.

Against such backgrounds, directed evolution of enzymes involved in PHA biosynthesis as well as metabolic engineering approaches of bacterial hosts will become the driving force to establish bioprocesses for the controlled production of PHAs with desired monomer compositions. In addition, systems-level analysis of metabolic, signaling, and regulatory networks is also making it possible to comprehensively understand global biological processes taking place in PHA-accumulating strains. The resultant knowledge will provide new targets and strategies for the improvement of PHA production, including tailor-made PHAs with desired monomer compositions and molecular masses.

Furthermore, from the viewpoint of preserving the ecosystem, bioplastics are most beneficial when they can be actually biodegraded. In order to achieve it, it is vital to elucidate the biodegradation mechanism of bioplastics and engineer their depolymerases. By contrast to PHA synthases, there have been very few protein engineering studies of PHA depolymerases using directed evolutionally methods, resulting in the less information about the improvement of PHA biodegradability as well as PHA depolymerases so far. In addition, as one of the recent trends in green polymer chemistry, *in vitro* bioplastic synthesis using isolated bioplastic-degrading enzymes has been developed because of the close relationship between the substrate specificities of the enzymes for polymer degradation and polymer synthesis. *In vitro* enzymatic polymerization offers many advantages, including easier control of polymer structure and monomer reactivity than conventional chemical methods. Accordingly, novel bioplastic-degrading enzymes evolved by protein engineering are expected to become useful biocatalysts for the bioplastic production in the future.

Here, we present the recent approaches of protein engineering with potential for a total recycle system of bioplastics via combination of biological production with biological degradation. In the future, custom-made prominent enzymes generated via evolutionary engineering will be utilized extensively to create high-performance bioplastics from renewable resources in various organisms and applied to effective and eco-friendly chemical recycling of bioplastics.

Author details

Tomohiro Hiraishi[1*] and Seiichi Taguchi[2,3]

*Address all correspondence to: thiraish@riken.jp

1 Bioengineering Laboratory, RIKEN Advanced Science Institute, Hirosawa, Wako-shi, Saitama, Japan

2 Division of Biotechnology and Macromolecular Chemistry, Graduate School of Engineering, Hokkaido University, Kita-ku, Sapporo, Japan

3 JST, CREST, Sanbancho, Chiyoda-ku, Tokyo, Japan

References

[1] Goldstein, M.C.; Rosenberg, M. & Cheng, L. (2013). Increased oceanic microplastic debris enhances oviposition in an endemic pelagic insect. *Biolog Lett*, published online. (DOI 10.1098/rsbl.2012.0298)

[2] Gross, R.A. & Kalra, B. (2002). Biodegradable polymers for the environment. *Science*, 297, 803-807.

[3] Hiraishi, T. & Taguchi, S. (2009). Enzyme-catalyzed synthesis and degradation of biopolymers. *Mini-Rev Org Chem*, 6, 44-54.

[4] Gomez, J.G.C.; Méndez, B.S.; Nikel, P.I.; Pettinari, M.J.; Prieto, M.A. & Silva, L.F. (2012). Making green polymers even greener: Towards sustainable production of polyhydroxyalkanoates from agroindustrial by-products, *Advances in Applied Biotechnology*, Marian Petre (Ed.), ISBN: 978-953-307-820-5, InTech.

[5] Tokiwa, Y.; Calabia, B.P.; Ugwu, C.U. & Aiba, S. (2009). Biodegradability of plastics. *Int J Mol Sci*, 10, 3722-3742.

[6]   Chen, G.Q. & Patel, M.K. (2012). Plastics derived from biological sources: Present and future: A technical and environmental review. *Chem Rev*, 112, 2082-2099.

[7]   Gasser, I.; Müller, H. & Berg, G. (2009). Ecology and characterization of polyhydroxy-alkanoate-producing microorganisms on and in plants. *FEMS Microbiol Ecol*, 70, 142-150.

[8]   Legat, A; Gruber, G.; Zangger, K.; Wanner, G. & Stan-Lotter, H. (2010). Identification of polyhydroxyalkanoates in *Halococcus* and other haloarchaeal species. *Appl Micro-biol Biotechnol*, 87, 1119-1127.

[9]   Jendrossek, D. & Handrick, R. (2002). Microbial degradation of polyhydroxyalka-noates. *Annu Rev Microbiol*, 56, 403-432.

[10] Kim, D.Y. & Rhee, Y.H. (2003). Biodegradation of microbial and synthetic polyesters by fungi. *Appl Microbiol Biotechnol*, 61, 300-308.

[11] Steinbüchel, A. & Valenthin, H.E. (1995). Diversity of bacterial polyhydroxyalkanoic acids. *FEMS Microbiol Lett*, 128, 219-228.

[12] Doi, Y. & Steinbüchel, A. Eds. *Biopolymers vol. 4; Polyesters III: Applications and Commercial Products*, Wiley-VCH, Weinheim, 2002.

[13] Verlinden, R.A.J.; Hill, D.J.; Kenward, M.A.; Williams, C.D. & Radecka, I. (2007). Bacterial synthesis of biodegradable polyhydroxyalkanoates. *J Appl Microbiol*, 102, 1437-1449.

[14] Chen, G.Q. (2009). A microbial polyhydroxyalkanoates (PHA) based bio- and materials industry. *Chem Soc Rev*, 38, 2434-2446.

[15] Auras, R.; Harte, B. & Selke, S. (2004). An overview of polylactides as packaging materials. *Macromol Biosci*, 4, 835-864.

[16] Tokiwa, Y. & Calabia, B.P. (2006). Biodegradability and biodegradation of poly(lactide). *Appl Microbiol Biotechnol*, 72, 244-251.

[17] Zhao, Y.M.; Wang, Z.Y.; Wang, J.; Mai, H.Z.; Yan, B. & Yang, F. (2004). Direct synthesis of poly(D, L-lactic acid) by melt polycondensation and its application in drug delivery. *J Appl Polym Sci*, 91, 2143-2150.

[18] Albertsson, A.C.; Edlund, U. & Stridsberg, K. (2000). Controlled ring-opening polymerization of lactones and lactides. *Macromol Symp*, 157, 39-46.

[19] Kricheldorf, H.R. (2001). Syntheses and application of polylactides. *Chemosphere*, 43, 49–54.

[20] Taguchi, S.; Yamada, M.; Matsumoto, K.; Tajima, K.; Satoh, Y.; Munekata, M.; Ohno, K.; Kohda, K.; Shimamura, T.; Kambe, H. & Obata, S. (2008). A microbial factory for lactate-based polyesters using a lactate-polymerizing enzyme. *Proc Natl Acad Sci USA*, 105, 17323-17327.

[21] Matsumoto, K. & Taguchi, S. (2010). Enzymatic and whole-cell synthesis of lactate-containing polyesters: toward the complete biological production of polylactate. *Appl Microbiol Biotechnol*, 85, 921-932.

[22] Shozui, F.; Matsumoto, K.; Motohashi, R.; Sun, J. A.; Satoh, T.; Kakuchi, T. & Taguchi, S. (2011). Biosynthesis of a lactate (LA)-based polyester with a 96 mol% LA fraction and its application to stereocomplex formation. *Polym Degrad Stabil*, 96, 499-504.

[23] Yamada, M.; Matsumoto, K.; Nakai, T. & Taguchi, S. (2009). Microbial production of lactate-enriched poly[(R)-lactate-*co*-(R)-3-hydroxybutyrate] with novel thermal properties. *Biomacromolecules*, 10, 677-681.

[24] Taguchi, S. (2010). Current advances in microbial cell factories for lactate-based polyesters driven by lactate-polymerizing enzymes: Towards the further creation of new LA-based polyesters. *Polym Degrad Stab*, 95, 1421-1428.

[25] Matsumura, S.; Mabuchi, K. & Toshima, K. (1997). Lipase-catalyzed ring-opening polymerization of lactide. *Macromol Rapid Commun*, 18, 477-482.

[26] Aldor, I.S. & Keasling, J.D. (2003). Process design for microbial plastic factories: metabolic engineering of polyhydroxyalkanoates. *Curr Opin Biotechnol*, 14, 475-483.

[27] Tokiwa, Y. & Jarerat, A. (2004). Biodegradation of poly(L-lactide). *Biotechnol Lett*, 26, 771-777.

[28] Kurcok, P.; Kowalczuk, M.; Hennek, K. & Jedlinski, Z. (1992). Anionic polymerization of β-lactones initiated with alkali-metal alkoxides: reinvestigation of the polymerization mechanism. *Macromolecules*, 25, 2017-2020.

[29] Fan, Y.; Nishida, H.; Shirai, Y. & Endo, T. (2003). Control of racemization for feedstock recycling of PLLA. *Green Chem*, 5, 575-579.

[30] Saeki, T.; Tsukegi, T.; Tsuji, H.; Daimon, H. & Fujie, K. (2005). Hydrolytic degradation of poly[(R)-3-hydroxybutyric acid] in the melt. *Polymer*, 46, 2157-2162.

[31] Abe, H. (2006). Thermal degradation of environmentally degradable poly(hydroxyalkanoic acid)s. *Macromol Biosci*, 6, 469-486.

[32] Schulze, B. & Wubbolts, M.G. (1999). Biocatalysis for industrial production of fine chemicals. *Curr Opin Biotechnol*, 10, 609-615.

[33] Ran, N.; Zhao, L.; Chen, Z. & Tao, J. (2008). Recent applications of biocatalysis in developing green chemistry for chemical synthesis at the industrial scale. *Green Chem*, 10, 361-372.

[34] Madison, L.L. & Huisman, G.W. (1999). Metabolic engineering of poly(3-hydroxyalkanoates): from DNA to plastic. *Microbiol Mol Biol Rev*, 63, 21-53.

[35] Rehm, B.H.A. (2003). Polyester synthases: natural catalysts for plastics. *Biochem J*, 376, 15-33.

[36] Taguchi, S. & Doi, Y. (2004). Evolution of polyhydroxyalkanoate (PHA) production system by "enzyme evolution": successful case studies of directed evolution. *Macromol Biosci*, 4, 146-156.

[37] Nomura, C.T. & Taguchi, S. (2007). PHA synthase engineering toward superbiocatalysts for custom-made biopolymers. *Appl Microbiol Biotechnol*, 73, 969-979.

[38] Taguchi, S.; Maehara, A.; Takase, K.; Nakahara, M.; Nakamura, H. & Doi, Y. (2001). Analysis of mutational effects of a polyhydroxybutyrate (PHB) polymerase on bacterial PHB accumulation using an in vivo assay system. *FEMS Microbiol Lett*, 198, 65-71.

[39] Taguchi, S.; Nakamura, H.; Hiraishi, T.; Yamato, I. & Doi, Y. (2002). *In vitro* evolution of a polyhydroxybutyrate synthase by intragenic suppression-type mutagenesis. *J Biochem*, 131, 801-806.

[40] Normi, Y.M.; Hiraishi, T.; Taguchi, S.; Abe, K.; Sudesh, K.; Najimudin, N. & Doi, Y. (2005). Characterization and properties of G4X mutants of *Ralstonia eutropha* PHA synthase for poly(3-hydroxybutyrate) biosynthesis in *Escherichia coli*. *Macromol Biosci*, 5, 197-206.

[41] Normi, Y. M.; Hiraishi, T.; Taguchi, S.; Sudesh, K.; Najimudin, N. & Doi, Y. (2005). Site-directed saturation mutagenesis at residue F420 and recombination with another beneficial mutation of *Ralstonia eutropha* polyhydroxyalkanoate synthase. *Biotechnol Lett*, 27, 705-712.

[42] Kichise, T.; Taguchi, S. & Doi, Y. (2002). Enhanced accumulation and changed monomer composition in polyhydroxyalkanoate (PHA) copolyester by in vitro evolution of *Aeromonas caviae* PHA synthase. *Appl Environ Microbiol*, 68, 2411-2419.

[43] Tsuge, T.; Watanabe, S.; Saito, S.; Hiraishi, T.; Abe, H.; Doi, Y. & Taguchi, S. (2007). Variation in copolymer composition and molecular mass of polyhydroxyalkanoate generated by saturation mutagenesis of *Aeromonas caviae* PHA synthase. *Macromol Biosci*, 7, 846-854.

[44] Amara, A.A.; Steinbüchel, A. & Rehm, B.H.A. (2002). In vivo evolution of the *Aeromonas punctata* polyhydroxyalkanoate (PHA) synthase: isolation and characterization of modified PHA synthases with enhanced activity. *Appl Microbiol Biotechnol*, 59, 477-482.

[45] Matsumoto, K.; Takase, K.; Yamamoto, Y.; Doi, Y. & Taguchi, S. (2009). Chimeric enzyme composed of polyhydroxyalkanoate (PHA) synthases from *Ralstonia eutropha* and *Aeromonas caviae* enhances production of PHAs in recombinant *Escherichia coli*. *Biomacromolecules*, 10, 682-685.

[46] Takase, K.; Taguchi, S. & Doi, Y. (2003). Enhanced synthesis of poly(3-hydroxybutyrate) in recombinant *Escherichia coli* by means of error-prone PCR mutagenesis, saturation mutagenesis, and *in vitro* recombination of the type II polyhydroxyalkanoate synthase gene. *J Biochem*, 133, 139-145.

[47] Takase, K.; Matsumoto, K.; Taguchi, S. & Doi, Y. (2004). Alteration of substrate chain-length specificity of type II synthase for polyhydroxyalkanoate biosynthesis by in vitro evolution: in vivo and in vitro enzyme assays. *Biomacromolecules*, 5, 480-485.

[48] Tsuge, T.; Saito, Y.; Narike, M.; Muneta, K.; Normi, Y.M.; Kikkawa, Y.; Hiraishi, T. & Doi, Y. (2004). Mutation effects of a conserved alanine (Ala510) in type I polyhydroxyalkanoate synthase from *Ralstonia eutropha* on polyester biosynthesis. *Macromol Biosci*, 4, 963-970.

[49] Matsumoto, K.; Takase, K.; Aoki, E.; Doi, Y. & Taguchi, S. (2005). Synergistic effects of Glu130Asp substitution in the type II polyhydroxyalkanoate (PHA) synthase: enhancement of PHA production and alteration of polymer molecular mass. *Biomacromolecules*, 6, 99-104.

[50] Matsumoto, K.; Aoki, E.; Takase, K.; Doi, Y. & Taguchi, S. (2006). In vivo and in vitro characterization of Ser477X mutations in polyhydroxyalkanoate (PHA) synthase 1 from *Pseudomonas* sp. 61-3: effects of beneficial mutations on enzymatic activity, substrate specificity, and molecular mass of PHA. *Biomacromolecules*, 7, 2436-2442.

[51] Shozui, F.; Matsumoto, K.; Sasaki, T. & Taguchi, S. (2009). Engineering of polyhydroxyalkanoate synthase by Ser477X/Gln481X saturation mutagenesis for efficient production of 3-hydroxybutyrate-based copolyesters. *Appl Microbiol Biotechnol*, 84, 1117-1124.

[52] Matsumoto, M.; Nagao, R.; Murata, T.; Arai, Y.; Kichise, T.; Nakashita, H.; Taguchi, S.; Shimada, H. & Doi, Y. (2005). Enhancement of poly(3-hydroxybutyrate-co-3-hydroxyvalerate) production in the transgenic *Arabidopsis thaliana* by the in vitro evolved highly active mutants of polyhydroxyalkanoate (PHA) synthase from *Aeromonas caviae*. *Biomacromolecules*, 6, 2126-2130.

[53] Matsumoto, K.; Murata, T.; Nagao, R.; Nomura, C.; Arai, S.; Takase, K.; Nakashita, H.; Taguchi, S. & Shimada, H. (2009). Production of short-chain-length/medium-chain-length polyhydroxyalkanoate (PHA) copolymer in the plastid of *Arabidopsis thaliana* using the evolved 3-ketoacyl-acyl carrier protein synthase III, *Biomacromolecules*, 10, 686-690.

[54] Solaiman, D.K. (2003). Biosynthesis of medium-chain-length poly(hydroxyalkanoates) with altered composition by mutant hybrid PHA synthases. *J Ind Microbiol Biotechnol*, 30, 322-326.

[55] Sheu, D.S. & Lee, C.Y. (2004). Altering the substrate specificity of polyhydroxyalkanoate synthase 1 derived from *Pseudomonas putida* GPo1 by localized semirandom mutagenesis. *J Bacteriol*, 186, 4177-4184.

[56] Niamsiri, N.; Delamarre, S.C.; Kim, Y.R. & Batt, C.A. (2004). Engineering of chimeric class II polyhydroxyalkanoate synthases. *Appl Environ Microbiol*, 70, 6789-6799.

[57] Valentin, H.E. & Steinbüchel, A. (1994). Application of enzymatically synthesized short-chain-length hydroxy fatty-acid coenzyme-a thioesters for assay of polyhydroxyalkanoic acid synthases. *Appl Microbiol Biotechnol*, 40, 699-709.

[58] Yuan, W.; Jia, Y.; Tian, J.M.; Snell, K.D.; Muh, U.; Sinskey, A.J.; Lambalot, R.H.; Walsh, C.T. & Stubbe, J. (2001). Class I and III polyhydroxyalkanoate synthases from *Ralstonia eutropha* and *Allochromatium vinosum*: characterization and substrate specificity studies. *Arch Biochem Biophys*, 394, 87-98.

[59] Tajima, K.; Satoh, Y.; Satoh, T.; Itoh, R.; Han, X.; Taguchi, S.; Kakuchi, T. & Munekata, M. (2009). Chemo-enzymatic synthesis of poly(lactate-*co*-(3-hydroxybutyrate)) by a lactate-polymerizing enzyme. *Macromolecules*, 42, 1985-1989.

[60] Yamada, M.; Matsumoto, K.; Shimizu, K.; Uramoto, S.; Nakai, T.; Shozui, F. & Taguchi, S. (2010). Adjustable mutations in lactate (LA)-polymerizing enzyme for the microbial production of LA-based polyesters with tailor-made monomer composition. *Biomacromolecules*, 11, 815-819.

[61] Han, X.; Satoh, Y.; Satoh, T.; Matsumoto, K.; Kakuchi, T.; Taguchi, S.; Dairi, T.; Munekata, M. & Tajima, K. (2011). Chemo-enzymatic synthesis of polyhydroxyalkanoate (PHA) incorporating 2-hydroxybutyrate by wild-type class I PHA synthase from *Ralstonia eutropha*. *Appl Microbiol Biotechnol*, 92, 509-517.

[62] Matsumoto, K.; Ishiyama, A.; Shiba, T. & Taguchi, S. (2011). Biosynthesis of glycolate-based polyesters containing medium-chain-length 3-hydroxyalkanoates in recombinant *Escherichia coli* expressing engineered polyhydroxyalkanoate synthase. *J Biotechnol*, 156, 214-217.

[63] Yang, T.H.; Jung, Y.K.; Kang, H.O.; Kim, T.W.; Park, S.J. & Lee, S.Y. (2011). Tailor-made type II *Pseudomonas* PHA synthases and their use for the biosynthesis of polylactic acid and its copolymer in recombinant *Escherichia coli*. *Appl Microbiol Biotechnol*, 90, 603-614.

[64] Tajima, K.; Han, X.; Satoh, Y.; Ishii, A.; Araki, Y.; Munekata, M. & Taguchi, S. (2012). In vitro synthesis of polyhydroxyalkanoate (PHA) incorporating lactate (LA) with a block sequence by using a newly engineered thermostable PHA synthase from *Pseudomonas* sp. SG4502 with acquired LA-polymerizing activity. *Appl Microbiol Biotechnol*, 94, 365-376.

[65] Ochi, A.; Matsumoto, K.; Ooba, T.; Sasaki, K.; Tsuge, T. & Taguchi, S. (2013). Engineering of class I lactate-polymerizing polyhydroxyalkanoate synthases from *Ralstonia eutropha* that synthesize lactate-based polyester with a block nature. *Appl Microbiol Biotechnol*, in press. (DOI 10.1007/s00253-012-4231-9)

[66] Knoll, M.; Hamm, T.M.; Wagner, F.; Martines, V. & Pleiss, J. (2009). The PHA Depolymerase Engineering Database: A systematic analysis tool for the diverse family of polyhydroxyalkanoate (PHA) depolymerases. *BMC Bioinform*, 10, 89-96.

[67] Miyazaki, S.; Takahashi, K.; Shiraki, M.; Saito, T.; Tezuka, Y. & Kasuya, K. (2002). Properties of a poly(3-hydroxybbutyrate) depolymerase from *Penicillium funiculosum*. *J Polym Environ*, 8, 175–182.

[68] Hisano, T.; Kasuya, K.; Tezuka, Y.; Ishii, N.; Kobayashi, T.; Shiraki, M.; Oroudjev, E.; Hansma, H.; Iwata, T.; Doi, Y.; Saito, T. & Miki, K. (2006). The crystal structure of polyhydroxybutyrate depolymerase from *Penicillium funiculosum* provides insights into the recognition and degradation of biopolyesters. *J Mol Biol*, 356, 993-1004.

[69] Handrick, R.; Reinhardt, S.; Focarete, M.L.; Scandola, M.; Adamus, G.; Kowalczuk, M. & Jendrossek, D. (2001). A new type of thermoalkalophilic hydrolase of *Paucimonas lemoignei* with high specificity for amorphous polyesters of short chain-length hydroxyalkanoic acids. *J Biol Chem*, 276, 36215-36224.

[70] Collins, T.; Gerday, C. & Feller, G. (2005). Chitinolytic enzymes: catalysis, substrate binding and their application. *FEMS Microbiol Rev*, 29, 3-23.

[71] Duo-Chuan, L. (2006). Review of fungal chitinases. *Mycopahologia*, 161, 345-360.

[72] Sukharnikov, L.O.; Cantwell, B.J.; Podar, M. & Shulin I.B. (2011). Cellulases: ambiguous nonhomologous enzymes in a genomic perspective. *Trends Biotechnol*, 29, 473-479.

[73] Beckham, G.T.; Ziyu Dai, Z.; James F Matthews, J.F. Momany, M.; Payne, C.M.; Adney, W. S.; Baker, S.E. & Himmel, M.E. (2012). Harnessing glycosylation to improve cellulase activity. *Curr Opin Biotechnol*, 23, 338-345.

[74] Behrends, A.; Klingbeil, B. & Jendrossek, D. (1996). Poly(3-hydroxybutyrate) depolymerases bind to their substrate by a C-terminal located substrate binding site. *FEMS Microbiol Lett*, 143, 191-194.

[75] Nojiri, M. & Saito, T. (1997). Structure and function of poly(3-hydroxybutyrate) depolymerase from *Alcaligenes faecalis* T1. *J Bacteriol*, 179, 6965-6970.

[76] Hiraishi, T.; Ohura, T.; Ito, S.; Kasuya, K. & Doi, Y. (2000). Function of the catalytic domain of poly(3-hydroxybutyrate) depolymerase from *Pseudomonas stutzeri*. *Biomacromolecules*, 1, 320-324.

[77] Kasuya, K.; Inoue, Y.; Tanaka, T.; Akehata, T.; Iwata, T.; Fukui, T. & Doi, Y. (1997). Biochemical and molecular characterization of the polyhydroxybutyrate depolymerase of *Comamonas acidovorans* YM1609, isolated from freshwater. *Appl Environ Microbiol*, 63, 4844-4852.

[78] Kasuya, K.; Ohura, T.; Masuda, K. & Doi, Y. (1999). Substrate and binding specificities of bacterial polyhydroxybutyrate depolymerases. *Int J Biol Macromol*, 24, 329-336.

[79] Shinomiya, M.; Iwata, T.; Kasuya, K. & Doi, Y. (1997). Cloning of the gene for poly(3-hydroxybutyric acid) depolymerase of *Comamonas testosteroni* and functional analysis of its substrate-binding domain. *FEMS Microbiol Lett*, 1997, 154, 89-94.

[80] Shinomiya, M.; Iwata, T. & Doi, Y. (1998). The adsorption of substrate-binding domain of PHB depolymerases to the surface of poly(3-hydroxybutyric acid). *Int J Biol Macromol*, 22, 129-135.

[81] Ohura, T.; Kasuya, K. & Doi, Y. (1999). Cloning and characterization of the polyhydroxybutyrate depolymerase gene of *Pseudomonas stutzeri* and analysis of the function of substrate-binding domains. *Appl Environ Microbiol*, 65, 189-197.

[82] Kemnitzer, J.E.; McCarthy, S.P. & Gross, R.A. (1992). Poly(β-hydroxybutyrate) stereoisomers: a model study of the effects of stereochemical and morphological variables on polymer biological degradability. *Macromolecules*, 25, 5927-5934.

[83] Jesudason, J. J.; Marchessault, R. H. & Saito, T. (1993). Enzymatic degradation of poly([R,S]β-hydroxybutyrate). *J Environ Polym Degrad*, 1, 89-98.

[84] Hocking, P. J. & Marchessault, R. H. (1993). Syndiotactic poly[(R,S)-.beta.-hydroxybutyrate] isolated from methyaluminoxane-catalyzed polymerization, *Polym Bull*, 30, 163-170.

[85] Abe, H.; Matsubara, I.; Doi, Y.; Hori, Y. & Yamaguchi, A. (1994). Physical properties and enzymatic degradability of poly(3-hydroxybutyrate) stereoisomers with different stereoregularities. *Macromolecules*, 27, 6018-6025.

[86] Hocking, P.J.; Timmins, M.R.; Scherer, T.M.; Fuller, R.C.; Lenz, R.W. & Marchessault, R.H. (1994). Enzymatic degradability of isotactic versus syndiotactic poly(β-hydroxybutyrate). *Macromol Rapid Commun*, 15, 447-452.

[87] Hocking, P.J.; Timmins, M.R.; Scherer, T.M.; Fuller, R.C.; Lenz, R.W. & Marchessault, R.H. (1995). Enzymatic degradability of poly(β-hydroxybutyrate) as a function of tacticity. *J Macromol Sci Pure Appl Chem A*, 32, 889-894.

[88] Timmins, M.R.; Lenz, R.W.; Hocking, P.J.; Marchessault, R.H. & Fuller, R.C. (1996). Effect of tacticity on enzymatic degradability of poly(β-hydroxybutyrate). *Macromol Chem Phys*, 197, 1193-1215.

[89] Abe, H. & Doi, Y. (1996). Enzymatic and environmental degradation of racemic poly(3-hydroxybutyric acid)s with different stereoregularities. *Macromolecules*, 29, 8683-8688.

[90] Bachmann, B. M. & Seebach, D. (1999). Investigation of the enzymatic cleavage of diastereomeric oligo(3-hydroxybutanoates) containing two to eight HB units. A model for the stereoselectivity of PHB depolymerase from *Alcaligenes faecalis* T1. *Macromolecules*, 32, 1777-1784.

[91] Scherer, T.M.; Fuller, R.C.; Goodwin, S. & Lenz, R.W. (2000). Enzymatic hydrolysis of oligomeric models of poly-3-hydroxybutyrate. *Biomacromolecules*, 1, 577-583.

[92] Kanesawa, Y.; Tanahashi, N.; Doi, Y. & Saito, T. (1994). Enzymatic degradation of microbial poly(3-hydroxyalkanoates). *Polym Degrad Stab*, 45, 179-185.

[93] Abe, H.; Doi, Y.; Aoki, H.; Akehata, T.; Hori, Y. & Yamaguchi, A. (1995). Physical properties and enzymic degradability of copolymers of (R)-3-hydroxybutyric and 6-hydroxyhexanoic acids. *Macromolecules*, 28, 7630-7637.

[94] Abe, H.; Doi, Y.; Hori, Y. & Hagiwara, T. (1997). Physical properties and enzymatic degradability of copolymers of (R)-3-hydroxybutyric acid and (S,S)-lactide. *Polymer*, 39, 59-67.

[95] Abe, H. & Doi, Y. (1999). Structural effects on enzymatic degradabilities for poly[(R)-3-hydroxybutyric acid] and its copolymers. *Int J Biol Macromol*, 25, 185-192.

[96] Tomasi, G.; Scandla, M.; Briese, B. H. & Jendrossek, D. (1996). Enzymatic degradation of bacterial poly(3-hydroxybutyrate) by a depolymerase from *Pseudomonas lemoignei*. *Macromolecules*, 29, 507-513.

[97] Abe, H.; Doi, Y.; Aoki, H. & Akehata, T. (1998). Solid-state structures and enzymatic degradabilities for melt-crystallized films of copolymers of (R)-3-hydroxybutyric acid with different hydroxyalkanoic Acids. *Macromolecules*, 31, 1791-1797.

[98] Kurokawa, K.; Yamashita, K.; Doi, Y. & Abe, H. (2006). Surface properties and enzymatic degradation of end-capped poly(l-lactide). *Polym Degrad Stab*, 91, 1300-1310.

[99] Kurokawa, K.; Yamashita, K.; Doi, Y. & Abe, H. (2008). Structural effects of terminal groups on nonenzymatic and enzymatic degradations of end-capped poly(L-lactide). *Biomacromolecules*, 9, 1071-1078.

[100] Yamashita, K.; Aoyagi, Y.; Abe, H. & Doi, Y. (2001). Analysis of adsorption function of polyhydroxybutyrate depolymerase from *Alcaligenes faecalis* T1 by using a quartz crystal microbalance. *Biomacromolecules*, 2, 25-28.

[101] Yamashita, K.; Funato, T.; Suzuki, Y.; Teramachi, S. & Doi, Y. (2003). Characteristic interactions between poly(hydroxybutyrate) depolymerase and poly[(R)-3-hydroxybutyrate] film studied by a quartz crystal microbalance. *Macromol Biosci*, 3, 694-702.

[102] Kikkawa, Y.; Yamashita, K.; Hiraishi, T.; Kanesato, M. & Doi, Y. (2005). Dynamic adsorption behavior of poly(3-hydroxybutyrate) depolymerase onto polyester surface investigated by QCM and AFM. *Biomacromolecules*, 6, 2084-2090.

[103] Kikkawa, Y.; Fujita, M.; Hiraishi, T.; Yoshimoto, M. & Doi, Y. (2004). Direct observation of poly(3-hydroxybutyrate) depolymerase adsorbed on polyester thin film by atomic force microscopy. *Biomacromolecules*, 5, 1642-1646.

[104] Murase, T.; Suzuki, Y.; Doi, Y. & Iwata, T. Nonhydrolytic fragmentation of a poly[(R)-3-hydroxybutyrate] single crystal revealed by use of a mutant of polyhydroxybutyrate depolymerase. *Biomacromolecules*, 3, 312-317.

[105] Fujita, M.; Kobori, Y.; Aoki, Y.; Matsumoto, N.; Abe, H.; Doi, Y. & Hiraishi, T. (2005). Interaction between PHB depolymerase and biodegradable polyesters by atomic force microscopy. *Langmuir*, 21, 11829-11835.

[106] Matsumoto, M.; Fujita, M.; Hiraishi, T.; Abe, H. & Maeda, M. (2008). Adsorption characteristics of P(3HB) depolymerase as evaluated by surface plasmon resonance and atomic force microscopy. *Biomacromolecules*, 9, 3201-3207.

[107] Shoseyov, O.; Shani, Z. & Levy, I. (2006). Carbohydrate binding modules: Biochemical properties and novel applications. *Microbiol Mol Biol Rev*, 70, 283-295.

[108] Papageorgiou, A.C.; Hermawan, S.; Singh, C.B. & Jendrossek, D. (2008). Structural basis of poly(3-hydroxybutyrate) hydrolysis by PhaZ7 depolymerase from *Paucimonas lemoignei*. *J Mol Biol*, 382, 1184-1194.

[109] Wakadkar, S.; Hermawan, S.; Jendrossek, D. & Papageorgiou, A.C. (2010). The structure of PhaZ7 at atomic (1.2 Å) resolution reveals details of the active site and suggests a substrate-binding mode. *Acta Cryst*, F66, 648–654.

[110] Bloom, J.D.; Meyer, M.M.; Meinhold, P.; Otey, C.R.; Macmillan, D. & Arnold, F.H. (2005). Evolving strategies for enzyme engineering. *Curr Opin Struc Biol*, 15, 447–452.

[111] Otten, L.G.; & Quax, W.J. (2005). Directed evolution: selecting today's biocatalysts. *Biomol Eng*, 22, 1–9.

[112] Johannes, T.W. & Zhao, H. (2006). Directed evolution of enzymes and biosynthetic pathways. *Curr Opin Microbiol*, 9, 261–267.

[113] Zhao, H.; Giver, L.; Shao, Z.; Affholter, J.A. & Arnold, F.H. (1998). Molecular evolution by staggered extension process (StEP) *in vitro* recombination. *Nat Biotechnol*, 16, 258–261.

[114] Lin, L.; Meng, X.; Liu, P.; Hong, Y.; Wu, G.; Huang, X.; Li, C.; Dong, J.; Xiao, L. & Liu, Z. (2009). Improved catalytic efficiency of endo-β-1,4-glucanase from *Bacillus subtilis* BME-15 by directed evolution. *Appl Microbiol Biotechnol*, 82, 671–679.

[115] Zhao, H.; Chockalingam, K. & Chen, Z. (2002). Directed evolution of enzymes and pathways for industrial biocatalysis. *Curr Opin Biotechnol*, 13, 104–110.

[116] Hiraishi, T.; Hirahara, Y.; Doi, Y.; Maeda, M. & Taguchi, S. (2006). Effects of mutations in the substrate-binding domain of poly[(R)-3-hydroxybutyrate] (PHB) depolymerase from *Ralstonia pickettii* T1 on PHB degradation. *Appl Environ Microbiol*, 72, 7331-7338.

[117] Hiraishi, T.; Komiya, N.; Matsumoto, N.; Abe, H.; Fujita, M. & Maeda, M. (2010). Degradation and adsorption characteristics of PHB depolymerase as revealed by kinetics of mutant enzymes with amino acid substitution in substrate-binding domain. *Biomacromolecules*, 11, 113-119.

[118] Hermawan, S. & Jendrossek, D. (2010). Tyrosine 105 of *Paucimonas lemoignei* PHB depolymerase PhaZ7 is essential for polymer binding. *Polym Degrad Stab*, 95, 1429-1435.

[119] Jendrossek, D., Schirmer, A. & Handrick, R. (1997). Recent advances in characterization of bacterial PHA depolymerase. In: Eggink, G.; Steinbuchel, A.; Poirier, Y. &

Witholt, B. editors. *1996 International symposium on bacterial polyhydroxyalkanoates*. Ottawa: NRC Research Press; p. 89-101.

[120] Ihssen, J.; Magnani, D.; Thöny-Meyer, L. & Ren, Q. (2009). Use of extracellular medium chain length polyhydroxyalkanoate depolymerase for targeted binding of proteins to artificial poly[(3-hydroxyoctanoate)-*co*-(3-hydroxyhexanoate)] granules. *Biomacromolecules*, 10, 1854-1864.

[121] Hiraishi, T.; Komiya, N. & Maeda, M. (2010). Y443F mutation in the substrate-binding domain of extracellular PHB depolymerase enhances its PHB adsorption and disruption abilities. *Polym Degrad Stab*, 95, 1370-1374.

[122] Kyte, J. & Doolittle, R.F. (1982). A simple method for displaying the hydropathic character of protein. *J Mol Biol*, 157, 105-132.

[123] Chou, P.Y. & Fasman,G.D. (1978). Prediction of the secondary structure of proteins from their amino acid sequence. *Adv Enzymol*, 47, 45-147.

[124] Zamyatnin A.A. (1972). Protein volume in solution. *Prog Biophys Mol Biol*, 24, 107-123.

[125] Lee, S.Y.; Choi, J.H. & Xu, Z. (2003). Microbial cell-surface display. *Trends Biotechnol*, 21, 45-52.

[126] Olsen, M.; Iverson, B. & Georgiou, G. (2000). High-throughput screening of enzyme libraries. *Curr Opin Biotechnol*, 11, 331–337.

[127] Samuelson, P.; Gunneriusson, E.; Nygren, P.Å. & Ståhl, S. (2002). Display of proteins on bacteria. *J Biotechnol*, 96, 129–154.

[128] Cote-Sierra J.; Jongert, E.; Gredan, A.; Gautam, D.C.; Parkhouse, M.; Cornelis, P.; De Baetselier, P. & Revets, H. (1998). A new membrane-bound OprI lipoprotein expression vector. High production of heterologous fusion proteins in Gram (-) bacteria and the implications for oral vaccination. *Gene*, 221, 25-34.

[129] Hiraishi, T.; Yamashita, K.; Sakono, M.; Nakanishi, J.; Tan, L.-T.; Sudesh, K.; Abe, H. & Maeda, M. (2012). Display of functionally active PHB depolymerase on *Escherichia coli* cell surface. *Macromol Biosci*, 12, 218-224.

[130] Tan, L.-T.; Hiraishi, T.; Sudesh, K. & Maeda, M. (2013). Directed evolution of poly[(*R*)-3-hydroxybutyrate] depolymerase using cell surface display system: functional importance of asparagine at position 285. *Appl Microbiol Biotechnol*, published online. (DOI 10.1007/s00253-012-4366-8)

[131] Zhao, H. & Zha, W. (2006). *In vitro* 'sexual' evolution through the PCR-based staggered extension process (StEP). *Nat Protoc*, 1, 1865–187.

Permissions

The contributors of this book come from diverse backgrounds, making this book a truly international effort. This book will bring forth new frontiers with its revolutionizing research information and detailed analysis of the nascent developments around the world.

We would like to thank Dr. Tomohisa Ogawa, for lending his expertise to make the book truly unique. He has played a crucial role in the development of this book. Without his invaluable contribution this book wouldn't have been possible. He has made vital efforts to compile up to date information on the varied aspects of this subject to make this book a valuable addition to the collection of many professionals and students.

This book was conceptualized with the vision of imparting up-to-date information and advanced data in this field. To ensure the same, a matchless editorial board was set up. Every individual on the board went through rigorous rounds of assessment to prove their worth. After which they invested a large part of their time researching and compiling the most relevant data for our readers. Conferences and sessions were held from time to time between the editorial board and the contributing authors to present the data in the most comprehensible form. The editorial team has worked tirelessly to provide valuable and valid information to help people across the globe.

Every chapter published in this book has been scrutinized by our experts. Their significance has been extensively debated. The topics covered herein carry significant findings which will fuel the growth of the discipline. They may even be implemented as practical applications or may be referred to as a beginning point for another development. Chapters in this book were first published by InTech; hereby published with permission under the Creative Commons Attribution License or equivalent.

The editorial board has been involved in producing this book since its inception. They have spent rigorous hours researching and exploring the diverse topics which have resulted in the successful publishing of this book. They have passed on their knowledge of decades through this book. To expedite this challenging task, the publisher supported the team at every step. A small team of assistant editors was also appointed to further simplify the editing procedure and attain best results for the readers.

Our editorial team has been hand-picked from every corner of the world. Their multi-ethnicity adds dynamic inputs to the discussions which result in innovative outcomes. These outcomes are then further discussed with the researchers and contributors who give their valuable feedback and opinion regarding the same. The feedback is then collaborated with the researches and they are edited in a comprehensive manner to aid the understanding of the subject.

Apart from the editorial board, the designing team has also invested a significant amount of their time in understanding the subject and creating the most relevant covers. They scrutinized every image to scout for the most suitable representation of the subject and create an appropriate cover for the book.

The publishing team has been involved in this book since its early stages. They were actively engaged in every process, be it collecting the data, connecting with the contributors or procuring relevant information. The team has been an ardent support to the editorial, designing and production team. Their endless efforts to recruit the best for this project, has resulted in the accomplishment of this book. They are a veteran in the field of academics and their pool of knowledge is as vast as their experience in printing. Their expertise and guidance has proved useful at every step. Their uncompromising quality standards have made this book an exceptional effort. Their encouragement from time to time has been an inspiration for everyone.

The publisher and the editorial board hope that this book will prove to be a valuable piece of knowledge for researchers, students, practitioners and scholars across the globe.

List of Contributors

Jingjing Li and Wei Han
Laboratory of Regeneromics, School of Pharmacology, Shanghai Jiao Tong University, Shanghai, China

Yan Yu
School of Agriculture and Biology, Shanghai Jiao Tong University, Shanghai, China

Alejandra Hernández-Santoyo
Instituto de Química, Universidad Nacional Autónoma de México, Mexico, D.F., Mexico

Aldo Yair Tenorio-Barajas, Victor Altuzar and Claudia Mendoza-Barrera
Laboratorio de Nanobiotecnología, Centro de Investigación en Micro y Nanotecnología, Universidad Veracruzana, Boca del Rio, Veracruz, Mexico

Héctor Vivanco-Cid
Instituto de Investigaciones Médico-Biológicas, Universidad Veracruzana, Boca del Río, Veracruz, Mexico

Francisco Valero
Department of Chemical Engineering, Engineering School, Universitat Autònoma de Barcelona, Bellaterra, Barcelona, Spain

Noriko Tabata, Kenichi Horisawa and Hiroshi Yanagawa
Department of Biosciences and Informatics, Faculty of Science and Technology, Keio University, Yokohama, Japan

Tomohisa Ogawa
Department of Biomolecular Science, Graduate School of Life Sciences, Tohoku University, Sendai, Japan
Nagahama Institute of Bio-Science and Technology, and Japan Science and Technology Agency, Japan

Tsuyoshi Shirai
Bioinfomatic Research Division, Nagahama, Shiga, Japan

Ari Rouhiainen and Heikki Rauvala
Neuroscience Center University of Helsinki, Finland

Helena Tukiainen and Pia Siljander
Department of Biosciences, University of Helsinki, Finland

Tomohiro Hiraishi
Bioengineering Laboratory, RIKEN Advanced Science Institute, Hirosawa, Wako-shi, Saitama, Japan

Seiichi Taguchi
Division of Biotechnology and Macromolecular Chemistry, Graduate School of Engineering, Hokkaido University, Kita-ku, Sapporo, Japan
JST, CREST, Sanbancho, Chiyoda-ku, Tokyo, Japan

Printed in the USA
CPSIA information can be obtained
at www.ICGtesting.com
JSHW011356221024
72173JS00003B/300